SPACE, TIME AND CAUSALITY

# SYNTHESE LIBRARY

STUDIES IN EPISTEMOLOGY,

LOGIC, METHODOLOGY, AND PHILOSOPHY OF SCIENCE



VOLUME 157

ROYAL INSTITUTE OF PHILOSOPHY CONFERENCES

VOLUME 1981

---

# SPACE, TIME AND CAUSALITY

*Edited by*

RICHARD SWINBURNE

*Department of Philosophy,*

*University of Keele, Keele, Staffordshire*

D. REIDEL PUBLISHING COMPANY

DORDRECHT : HOLLAND / BOSTON : U.S.A.

LONDON : ENGLAND

**Library of Congress Cataloging in Publication Data**
Main entry under title:

Space, time, and causality.

(Synthese library ; v. 157)
'The papers in this volume are revised versions of papers read to a conference at the University of Keele in September 1981, sponsored by the Royal Institute of Philosophy'–Pref.
Bibliography: p.
Includes index.
1. Space and time–Congresses. 2. Causation–Congresses.
I. Swinburne, Richard. II. Royal Institute of Philosophy.
BD632.S63 1982 110 82-18123
ISBN 90-277-1437-1

---

Published by D. Reidel Publishing Company,
P.O. Box 17, 3300 AA Dordrecht, Holland.

Sold and distributed in the U.S.A. and Canada
by Kluwer Boston Inc.,
190 Old Derby Street, Hingham, MA 02043, U.S.A.

In all other countries, sold and distributed
by Kluwer Academic Publishers Group,
P.O. Box 322, 3300 AH Dordrecht, Holland.

D. Reidel Publishing Company is a member of the Kluwer Group.



Printed in The Netherlands

## TABLE OF CONTENTS

## CAUSALITY, RELATIVITY, AND THE EINSTEIN–PODOLSKY–ROSEN PARADOX

# FOREWORD

The Royal Institute of Philosophy has been sponsoring conferences in alternate years since 1969. These have from the start been intended to be of interest to persons who are not philosophers by profession. They have mainly focused on interdisciplinary areas such as the philosophies of psychology, education and the social sciences. The volumes arising from these conferences have included discussions between philosophers and distinguished practitioners of other disciplines relevant to the chosen topic.

Beginning with the 1979 conference on 'Law, Morality and Rights' and the 1981 conference on 'Space, Time and Causality' these volumes are now constituted as a series. It is hoped that this series will contribute to advancing philosophical understanding at the frontiers of philosophy and areas of interest to non-philosophers. It is hoped that it will do so by writing which reduces technicalities as much as the subject-matter permits. In this way the series is intended to demonstrate that philosophy can be clear and worthwhile in itself and at the same time relevant to the interests of lay people.

STUART BROWN
*Honorary Assistant Director*
*Royal Institute of Philosophy*

# PREFACE

The papers in this volume are revised versions of papers read to a conference at the University of Keele in September 1981, sponsored by the Royal Institute of Philosophy. The general theme of the conference was the implications for our understanding of space, time and causation of the developments of modern physics, and especially of those two great theories of modern physics – Relativity Theory and Quantum Theory.

It is very sad to have to record that after participating in the conference with his usual vigour, but before he could complete the final revisions to his paper, J. L. Mackie died on December 12th 1981. I should like to dedicate this volume to his memory. John Mackie believed that philosophical argument, applied to knowledge provided by other disciplines and above all by science, could lead to a clear and unified world-view. He has contributed in a major way to debates about most areas of philosophy. His enthusiasm for the subject showed in his readiness to argue about it with almost any opponent, however humble or awkward. He always did so with great courtesy; as one who argued with him about many philosophical matters from a diametrically opposed position over many years can certainly testify. Oddly enough, the issue discussed in our two contributions to the present volume was one of very few about which we had a small measure of agreement.

*January 1982* RICHARD SWINBURNE

# INTRODUCTION

A major role of the philosopher of science is to put major scientific theories under the philosophical microscope. Once the scientist has constructed his theory, given it precise mathematical form, used it to make predictions, and found that these confirm the theory, the philosopher comes and examines the theory. He asks first what does the theory mean, what is it telling us about the world. Some philosophers and scientists (variously called instrumentalists, operationalists, or anti-realists) have held that the whole meaning of the theory is contained in its predictions about what can be observed – e.g. that the atomic theory of chemistry, as put forward in the nineteenth century, was simply a shorthand summary of observable data, of which chemical substances combine with other chemical substances in what proportions by mass and volume to form other substances; of the viscosity, boiling points and melting points of substances. But most philosophers and scientists these days are realists. They think of scientific theories not just as predicting devices, but as devices which tell us about the unobservable entities and properties which lie behind and cause the phenomena which we observe. They understand talk about 'molecules', 'atoms', 'protons', 'photons', 'electrons' and 'neutrinos' as talk about very small things which possess properties such as mass and charge, not just as terms in a scientific theory which is a successful predicting device. Yet even if you adopt this realist attitude to science, it is not always clear just what the scientific theory is claiming – are electrons supposed to be small round material objects, like billiard balls but very much smaller; or are they supposed to be rather different from anything observable on the large scale? If the world about which the scientist tries to tell us is very different from the world which we can observe with the naked eye, he will have a problem in describing it – for our ordinary language is designed for talk about the observable world. The scientist will therefore need to use words in slightly different senses from their normal senses. The role of the philosopher is to show when and how this is happening, and so to make clear just what the scientific theory is claiming. The physicist may tell us that electrons may be regarded either as waves or as particles. But if he is claiming that they are both waves and particles, the philosopher needs to clarify this claim. For superficially it is hard to make sense of this – how can something be both the propagation of a disturbance in a medium (which is what a wave is) and a material object? A second task for the philosopher when he has spelled out what the theory means is to show why the scientist's evidence is in fact evidence for that interpretation of the theory. The theory tells us about things unobservable. Why should we take particular observations as evidence in favour of the existence of just those things?

Relativity Theory and Quantum Theory present abundant scope for philosophical investigation. They are both theories of great mathematical sophistication which have proved very successful predictors of observations. But the descriptions in words which physicists give of the meaning of these theories are highly paradoxical. Ordinarily we suppose that an event occurs at a definite time and place – i.e. at a definite interval after some other specified event and at a definite distance from it. But the Special and General Theories of Relativity, at any rate as they are normally expounded, challenge this understanding. We are told that the distance in space and the interval in time between events depends on the frame of reference in which rulers and clocks used for measurement are situated – the Earth, say, or a spaceship moving across the Earth's surface. There is no true distance in space or interval in time between events; there is just 'distance in $F$', and 'distance in $F'$', 'temporal interval in $F$', and 'temporal interval in $F'$'. Again, we ordinarily wish to distinguish between real motion (e.g. the motion of a train across the countryside) and apparent motion (e.g. the motion of the countryside as seen from the train). But relativity theory seems to tell us that in reality there is no such thing as absolute motion. Indeed it seems to go even further and say that space and time are not separate things in which there are separate intervals. The real entity is space-time, intervals in which really exist – any division of these intervals into spatial and temporal components is an arbitrary one. We need to elucidate and make sense of these claims, if we can; and clarify precisely whether the evidence of observation supports these claims.

Similar problems arise with Quantum Theory. Ordinarily we suppose that physical objects are at definite places at definite times moving with definite velocities and momenta. Yet Quantum Theory, at any rate on some interpretations, seem to suggest not merely that we cannot know the exact position and momentum of a particle, such as an electron or photon, but that it does not have these. It only has the probability of being in a certain place and having a certain momentum – until it is 'observed' (i.e. interacts with some measuring apparatus which gives a reading which we can observe). Then if its position is measured, this acquires a definite value, while its momentum becomes very imprecise (i.e. there is only a certain probability with respect to any small range of momenta that its momentum lies within that range); and if the momentum is measured, it acquires a definite value, while position becomes very imprecise. All this needs clarifying. How can an electron have an imprecise position, except in the sense that we do not know what its position is? How can measurement cause it to have a position which it did not have before?

Also, Quantum Theory is a statistical theory in the sense that it only allows prediction of values of measurements with varying degrees of probability, does not allow us to predict the position of a photon or the moment of decay of an atom with certainty. Yet does Quantum Theory simply tell us that there is a limit to what we can discover about the world, or does it tell us that nature is really indeterministic – that nothing causes the photon to occupy the position it does or the atom to decay at the moment it does?

So Relativity Theory and Quantum Theory call into question many of our ordinary assumptions about space, time, and causality. Other modern scientific theories are relevant too. Thermodynamics introduced the valuable concept of entropy. The entropy of a system increases in so far as the different bodies of the system become closer in temperature to each other. The second law of thermodynamics says the entropy of a closed system (one isolated from outside influences) always increases. It has been suggested that the difference between an earlier state of a system and a later state is just this difference between the system being in a lower state and being in a higher state of entropy; that the direction of time just is the direction of increase of entropy. But is there not more to earlier and later than that?

It is to topics within this general area that the papers in this volume are devoted. The paper by Mackie begins by distinguishing various different issues which are at stake in a controversy about whether space and time are absolute – e.g. there is one issue whether places can be reidentified in a non-frame-relative way, and so whether we can distinguish relative from absolute motion; it is another issue whether space is a thing separate from the objects which occupy it, and so on. He makes out a case for absolutism on many of the issues which he distinguishes. He claims that the Special Theory of Relativity does not rule out the existence of an absolute frame of reference. We may not be able to discover when a body is absolutely at rest, or absolutely at such and such a distance from another body; but sometimes bodies are at rest or at absolute distances from each other. There are truths here, he urges, even if we cannot discover them. Dorling argues that Mackie has made quite unjustifiable steps beyond the so well established Theory of Relativity.

Zahar claims that Dorling's criticism suffers from unjustifiable verificationist and operationalist prejudices. It is meaningful to claim that there is an absolute rest-frame, even if we cannot discover it. However, he claims, against Mackie, that it is most unlikely that there is such a frame. If there were such a frame then, given the well justified predictions of the theory of Relativity that the laws of nature acquire the same form when measurements are made relative to any inertial frame, Nature would be engaged in a huge 'conspiracy of slience' to prevent us detecting the rest frame; and that is most unlikely.

A theme running through these discussions is reductionism – the suggestion that time and space or Space-Time are really something else. Sklar disringuishes two different kinds of attempt to reduce Space-Time to something else – e.g. to reduce talk of events being before or after other events to talk about states of entropy, or talk about events being causes or effects. Causes precede their effects, and so, the latter suggestion goes, an event $E_2$ is after another event $E_1$ if $E_1$ could cause an event (e.g. by sending a signal) 'locally simultaneous' with (i.e. at the same place and time as) $E_2$. This 'Robbian' programme attempts to reduce all talk of before, after and simultaneous with (for events at a distance from each other) to talk of causal connectibility (or continuity) and 'local simultancity'. Sklar distinguishes two kinds of philosophical motivation for such reductionist programmes. In my response to Sklar, I concentrate on one of these kinds of motivation – the philosophical doctrine of verificationism which lay behind the Robbian programme, and to which reference is made in a number of papers in the volume. The verificationist holds that the only sentences which have a truth value are ones which in some sense can be tested, are 'verifiable' or 'falsifiable' by observation. I cast some doubt on verificationism, and argue that the only plausible form of verificationism does not support a Robbian programme.

Causes, I have noted, precede their effects. Perhaps sometimes they do not, but in general the direction of causation is the direction of time. But is it just a matter of definition that causes precede their effects (i.e. we would not count something as a 'cause' unless it was earlier than its suggested effect)? Or is there some deeper reason – some necessity in the world arising from the natures of time and cause which make their directions coincide? These are the topics discussed by Healey and Newton-Smith.

Philosophy of science merges at one end into more general philosophy, at the other end into science itself. My own contribution to this volume lies at the former end of the spectrum; Cartwright's lies at the other. Her concern is with the fact that the normal interpretation of Quantum Theory gives a very special place to measurement. Measurement seems to make a difference to what happens. The Schrodinger equation of Quantum Theory predicts, for example, deterministically, how a light wave, 'associated with' a photon, the particle of light, will pass through a slit and move towards a photographic plate. But when it interacts with the plate, the photon makes a mark on the plate which reveals its location. The interaction is thus a process of measurement. The wave, as it were, collapses on to a point and this phenomenon is known as the 'reduction of the wave packet'. This process, which is indeterministic, is not governed by the Schrödinger equation; yet surely it ought to be, for measurement is just interaction with another physical object. Cartwright suggests an amendment to Quantum Theory which will allow it to represent the two kinds of evolution of systems, deterministic and indeterministic, as special cases of a more general kind of evolution. In reply Butterfield points out that Cartwright's programme is only a programme and needs detailed development and experimental confirmation; and he adds certain general comments on her philosophy of science.

One particular difficulty in understanding Quantum Theory is raised by the Einstein-Podolski-Rosen paradox. This seems to suggest that measuring a variable at one location instantaneously causes a variable at another location to have a certain value. But there does not seem to be any process known to physics by which this can happen, and anyway it seems to violate the apparent thesis of Relativity Theory that no effect can be propagated with a velocity faster than that of light. Redhead analyses the extent to which the EPR paradox shows that Quantum Theory does allow propagation of effects with velocities greater than that of light. He then goes on to consider whether, if it did allow this, Quantum Theory would thereby be in conflict with Relativity Theory. He claims that it would not: that there are ways of understanding Relativity Theory which do not rule out effects being produced at velocities greater than that of light – as would happen if there were 'tachyons', particles with such a velocity. Gibbins argues that there is no need to postulate super-luminal propagation of effects, in order to resolve the difficulties raised by the EPR paradox. The trouble arises, he claims, from supposing that our normal ('classical') logic holds on the subatomic level. If we use instead a new Quantum Logic, paradoxes vanish.

The papers, it will be seen, pick out only certain specific issues within the general area of philosophical problems about space, time, and causality raised by modern physical theories. In the conference debates the main clash seemed to be between those who took physicists' interpretations of their theories as seriously as they took the formulae and predictions of those theories, and those who felt that physicists needed help from philosophers about the ontological commitments of their theories. The former group felt that the formulae of a theory and their interpretation go naturally together (for the physicist has produced them together as an integral theory). Hence if a physicist interprets the Lorentz formula of the Special Theory of Relativity showing how the value of '*t*' varies with the frame of reference in which it is measured, as saying that the temporal interval between two events depends on the frame and so whether two events are simultaneous or not depends on the frame, we should take his word for it. The latter group felt that philosophers are better with words than are physicists. We need to inquire whether the physicist's verbal translation of '*t*' as 'temporal interval' is correct. Physicists need to have their attention drawn to the meaning and logical consequences of what they are saying if they are using words in ordinary senses, and to possible contradictions in what they say. If physicists claim to be using words in novel senses as they reasonably may, they need to explain what those senses are and how they are related to ordinary senses; and perhaps philosophers can help in that elucidation.

# ABSOLUTE *VERSUS* RELATIVE SPACE AND TIME

J. L. MACKIE *

# THREE STEPS TOWARDS ABSOLUTISM

## 1. INTRODUCTION: ISSUES AND GENERAL ARGUMENTS

Quite a number of philosophers of science have argued, in recent years, for at least some kind of absolutism about space or time or space-time; but most philosophers who do not specialize in this area seem to take a relativist view, and indeed a fairly extreme form of relativism, to be simply obvious, or to be established beyond the need for controversy. This paper is addressed primarily to such general philosophers, and its purpose is at least to disturb their complacency.

There are many issues, not just one, that come under the heading of 'absolutism *versus* relativism about space and time', and the first task is to distinguish them. One question is whether space and time are entities, existing independently of things and processes, whether they are proper objects of reference, or whether we should speak only of spatial and temporal features – qualities and relations – of things and processes. This is the issue about absolute or relative existence. But we might decide that space and time are not altogether distinct from one another, that we have rather space-time which can be sorted out into spatial and temporal dimensions only relatively to some arbitrarily chosen frame of reference; even so, there will still be an issue about the absolute or relative existence of space-time, the question whether it is a proper object of reference, whether it exists independently or only as a collection of features of things and processes. Distinct from these issues of existence is the issue about position: are there absolute positions in space and dates in time (or, again, points in space-time), or does one thing or event have a position or a date or a space-time location only relatively to some other thing(s) or event(s)? Again, is there an absolute difference between motion and rest, or can one thing move or not move only in relation

* [The paper printed here consists of the paper read to the Keele conference, with a number of minor corrections which Mr. Mackie made before his death. He was also intending to write an additional note, in part commenting on some of the objections made in the course of the discussion at the conference, but the note was never written. I am most grateful to Mrs Joan Mackie for considerable help in discovering the corrections to the paper. – *Ed.*]



*R. Swinburne (ed.), Space, Time and Causality*, 3–22.

to another? Similarly, is there an absolute difference between acceleration – change of motion, including rotation – and non-acceleration, or is there change or non-change of motion only relatively to some arbitrarily chosen frame of reference? There are further issues about metrical features. One kind of absolutism will say that things have intrinsic spatial lengths, or that there are intrinsic metrical relations of distance between spatial positions – whether such positions themselves are absolute or merely relative – and similarly that processes have intrinsic temporal lengths or durations, or that there are intrinsic metrical relations of time-interval between occurrences – whether these occurrences in themselves have absolute or merely relative dates, positions in time. The relativism opposed to this kind of absolutism will say that all such metrical features are merely relative, that the most we have is that one thing or process or interval is equal or unequal in spatial or temporal length to another. But this kind of relativism still leaves room for a more limited kind of absolutism, namely the thesis that there are absolute equalities and inequalities of spatial length and distance and of temporal interval and duration. Opposed to this kind of absolutism is the relativism which holds that even these metrical relations are themselves relative to some arbitrarily chosen metrical system or method of measurement, or to some 'observer' – for example, that equalities or inequalities of temporal duration hold only in relation to some chosen clock or set of clocks. And even this does not exhaust the range of questions. For it can be at least plausibly argued that even if the measures of space and time separately, the equalities of length or distance on its own and the equalities of duration on its own, are thus relative to clocks or space-measuring procedures and instruments, yet there are intrinsic equalities and inequalities of time-like length and space-like length along space-time paths, so that, as we may put it, space-time has an intrinsic metric even if space and time separately do not: this would be a further kind of absolutism, while the denial of any such intrinsic space-time metric would be a yet more radical relativism. In fact, when we set them out systematically we find that there are at least fourteen distinct issues, fourteen possible absolutist theses (as shown in Table I), with a relativist thesis as the denial of each of these.

But of course these issues are not wholly independent of one another. Roughly speaking, as we go down the table we come to progressively weaker absolutist theses – and by contrast to stronger and stronger or more and more radical relativist theses. Equally, the conjunction of the separate theses about space and time on any one line will entail the corresponding thesis about space-time on the same line, but neither of those separate theses will

TABLE I
Varieties of absolutism and relativism

There may be absolutisms and contrasting relativisms about:

| | | |
|---|---|---|
| 1.1. the *existence* of space | 1.2. of time | 1.3. of space-time |
| 2.1. *position* in space | 2.2. in time | 2.3. in space-time |
| | 3. *motion/rest* | |
| | 4. *acceleration/non-acceleration* | |
| 5.1. *length/distance* in space | 5.2. *duration/interval* in time | 5.3. *time-like and space-like lengths* of paths in space-time |
| 6.1. *equality/inequality of length/distance* in space | 6.2. *equality/inequality of duration/interval* in time | 6.3. *equality/inequality* of *time-like and space-like lengths* of paths in space-time |

be entailed by the corresponding space-time thesis. For example, absolutism about spatial positions and dates requires, but is not required by, absolutism about motion, and absolutism about motion requires but is not required by absolutism about acceleration; if space-time paths have intrinsic lengths, then there will be absolute equalities and inequalities between them, but there might be such absolute equalities and inequalities even if they had no intrinsic lengths. If space and time each exist as independent entities, we should at least expect this fact to carry with it all the other absolutisms – though they *might* perhaps be amorphous entities, with no metrical features or even with no distinguishable parts – but the various absolutisms lower down the table might hold without such absolute existence; and the same applies to absolutism about the existence of space-time and the various absolutisms below this in the right-hand column.

There is a connection of another kind between these issues. In each case the relativist view is commonly supported by the same general line of argument. We cannot observe space or time by itself apart from things or processes, so we have no right to assert its existence as an independent entity. We cannot fix a thing's position except in relation to other things. There would be no detectable difference if everything in the universe were a mile further north than it is, so this form of words fails to specify any real difference. We cannot tell whether one thing is moving or not on its own, but only whether it is moving in relation to other things or to ourselves. There would be no way of detecting whether the whole universe was moving together. And so on. The relativist typically argues that whatever the particular absolutism which he is opposing asserts is unobservable, or at any rate not directly observable, and therefore cannot be real.

But how does the alleged unobservability bear upon the issue in each case? The relativist must be relying not on the unobservability alone but also on some philosophical principle which authorizes him to deny or reject or dismiss whatever cannot be (directly) observed. In fact, several such principles have been put forward.

One of these is Leibniz's principle of sufficient reason: nothing happens without a sufficient reason, and in particular God does nothing without a sufficient reason. But, Leibniz argues, there could be no sufficient reason why everything in the universe should be where it is rather than all together being a mile further north; so there can be no real difference between these alternatives, as absolutism about spatial position holds that there is. If there were absolute positions in time, God would have been faced with the choice whether to create the world at one time or, say, twenty-four hours later, letting it run on in just the same way. But since the difference between these would be utterly undetectable for anyone within the universe, and neither course of events could be better than the other in any way, he could not have had a sufficient reason for preferring one to the other. But if there are no absolute temporal positions, God is spared the embarrassment of such an unresolvable choice. In this way the principle of sufficient reason would support various relativisms. However, there is no sufficient reason why we should accept this principle.

Another suitable principle is the verificationist theory of meaning. If the meaning of every statement is constituted by the method(s) by which it would be verified, a supposed statement which was utterly cut off from the possibility of verification would lack meaning, and so would not really be a statement after all. So if absolute motion, for example, is undetectable, the statement that something is moving absolutely, or that it is absolutely at rest, must after all say nothing.

Even if this principle is accepted, it seems to leave room for a defence of at least some varieties of absolutism. For example, it has been argued, by Newton and by others following him, that although absolute acceleration is not directly observable, it is indirectly observable and indeed measurable, by way of its dynamic causes and effects. For example, if there were two equal metal spheres joined by a spring, then if we knew the masses of the spheres and the elastic characteristics of the spring we could determine whether, and if so at what angular velocity, this little system was rotating absolutely merely by seeing whether, and if so how much, the spring was extended.

But the verificationist-relativist will reply that if absolute acceleration is never directly observed, we cannot infer it from such effects – or, likewise, from any causes. We cannot establish or even confirm the causal law on which such an inference must rely unless we directly observe, sometimes at least, each of the terms that the law connects. And in default of such an inference, if we claim that, for example, the extension of the spring is a measure of absolute rotation, we must simply be introducing the name 'abolute rotation' as a name for such a directly observed change: 'The system is rotating absolutely with such and such an angular velocity' can only mean 'The spring is extended so far'. And in general instead of being able thus to introduce indirectly observed items, we succeed only in introducing what are likely to be misleading names for the features which are directly observed, and which the absolutist was trying to use as evidence for something else.

However, I shall show (in Section 2) that this reply cannot be sustained. The criticism of it there will bring out the weakness of the verificationist theory of meaning; but this theory has in any case some very unpalatable cosequences in both scientific and commonsense contexts, and it can, I believe, be decisively refuted.[1]

Another principle on which the appeal to unobservability may rely is that of economy of postulation, Ockham's razor. An entity, or a quality, or a relation, that is never directly observed is one that we are not forced to admit, and therefore one that we could do without. However, there is more than one kind of economy, and I shall show (in Section 3) that it may be more economical in an important sense to postulate at least some of the items favoured by absolutism than to do without them as the relativist would prefer.

A fourth principle is the operationalist one, that science should use only operationally defined terms. But this would at most exclude absolutist terms and theses from science: it would not settle any philosophical issues about the existence or non-existence of absolute space or time or space-time, or about the absolute or relative status of the various other features. In any case operationalism derives whatever plausibility it has as a programme for science either from a verificationist theory of meaning or from some principle of the economy of postulation. If these fail, it has no independent appeal.

I maintain, therefore, that there is no cogent general argument for the diferent varieties of spatio-temporal relativisms, based on the impossibility of observing directly each of the controversial absolute entities or features. Rather, the specific issues have to be examined one by one, to see whether, in each case, the absolutist has good grounds for postulating the item(s) in question. *That is, abslutisms cannot be systematically ruled out: nevertheless each particular absolutism will need to have a case developed for it before it can be ruled in.*

Obviously, I cannot hope to deal with all these issues in one paper; instead, in the three following sections, I shall discuss a few of them, but I hope, in doing this, to illustrate some general principles which should guide us in this area as a whole.

## 2. ABSOLUTE ACCELERATION

I shall treat the problem of absolute acceleration fairly summarily, because it has been thoroughly discussed by (for example) Newton, Mach, Reichenbach, and Swinburne, and most of what needs to be said about it emerges directly from their discussions.[2]

As everyone knows, Newton argued that since, if we rotate a bucket of water, we find that the surface becomes concave – and the more concave the more rapidly it rotates – we have here an indication of absolute acceleration: the 'centrifugal force' which seems to push the water outwards so that it builds up at the sides of the bucket is really a symptom of the fact that the water is constantly being forced to accelerate towards the centre of the bucket. Similarly, with the two spheres joined by a spring, the tension of the spring is needed to exert on each sphere the force which makes it accelerate towards the centre as the system rotates. The result of the bucket experiment can be checked by anyone in his back garden; observations that are practically equivalent to the two-sphere experiment can be made on planetary and satellite systems and double stars. So far, these results favour Newton: there is independent evidence for the presence of forces which are just the ones that we should expect to find if there were absolute rotations and therefore absolute accelerations involved in them. Moreover, these rotations are not in general rotations relative to some nearby body. The bucket's rotation is indeed relative to the earth, but there is nothing corresponding to this in the other cases. But, as Mach points out, the rotations that are doing the work need not be interpreted as absolute rotations. For they are, obviously, rotations relative to 'the great masses of the universe', the fixed stars and beyond them the galaxies. However, Mach erred in arguing *a priori* that it must be this relative rotation that is doing the work. As

Reichenbach points out, it is an empirical question whether this is so. Suppose that we find centrifugal phenomena on an earth $E_1$ which is rotating relatively to a surrounding shell of fixed stars $F_1$; then if there were, far away, another earth $E_2$ stationary relative to $F_1$, but with a shell of stars surrounding it, $F_2$, which is stationary relative to $E_1$, then $E_1$ and $E_2$ will each have the same rotation relative to its own surrounding shell. But it is an empirical question whether the same centrifugal phenomena are in fact found on both. If they are, this will confirm Mach's hypothesis: since the effects are correlated with the relative rotations between each earth and its own star shell, it is these relative rotations that are doing the work. But if the centrifugal phenomena are found in $E_1$ but not in $E_2$, Mach's hypothesis is disconfirmed. We may be tempted to say that in this case Newton's interpretation, that it is an absolute rotation of $E_1$ that is producing these effects, is confirmed; but Reichenbach argues that this would be too hasty a conclusion. *Some* absolute rotation, he concedes, is responsible; but it might be either an absolute rotation of $E_1$ or an absolute rotation of tis surrounding shell $F_1$. And Reichenbach thinks that we shall be unable to decide which it is. But, as Swinburne says, this is not so: some further additions to the possible observations could decide this. If we have other earths, $E_3$ and $E_4$ and so on, with various different rotations relatively to $E_1$ but within the same star shell $F_1$, we shall be able to decide, by seeing what centrifugal effects, if any, they exhibit, whether these effects are correlated with what would have to be the different absolute rotations of each earth or with the rotation of the common surrounding star shell $F_1$. The whole discussion is a straightforward application of the methods of eliminative induction that are known as Mill's Methods. With a sufficient range of partly similar and partly different situations, we could show that various proposed factors in turn are not causally responsible for the 'centrifugal effects'; and the elimination of enough rival hypotheses could in the end powerfully confirm Newton's view that these effects are being produced by an absolute rotation of the body in which these effects appear, and hence by an absolute acceleration of its parts. Of course, the empirical outcome of such a range of observations could go the other way, and confirm, say, Mach's relativist interpretation. All I am insisting on at present is that it is an empirical question; I am protesting against the tendency of many relativists to suppose that the relativist account is the only possible one, that it can be established on some general philosophical grounds without waiting for the empirical evidence.

But now we have a puzzle on our hands. How does this manage to survive as an empirical question? How does it escape the general argument sketched in Section 1 against the possibility of introducing items which are never directly observed, for which we have at best indirect evidence? Specifically, how can we first give meaning to a statement of the form '*X* has such-and-such an absolute rotation' and secondly confirm it? If the very notion of rotation is first introduced in connection with the observed rotation of one thing in relation to another, how can we even frame, without internal contradiction, the concept of a rotation that is not relative to anything?

Well, we can first introduce, quite arbitrarily, the notion of a purely abstract standard or frame of reference. There is a frame of reference, for example, with respect to which this book is rotating clockwise as seen from above about an axis perpendicular to its cover through its centre of gravity at sixty revolutions a minute; and there is another frame of reference with respect to which the same book is rotating about the same axis in the opposite direction at twice that angular velocity. Abstract frames of reference are cheap: you can have as many of them as you like at no cost at all; and as such they have no physical significance. But suppose that there is some one abstract frame of reference such that certain dynamic causes and effects are systematically associated with rotations (which vary in magnitude, in their axes of rotation, and in direction) relative to it. Suppose, that is, that there is one 'preferred' frame of reference rotations relative to which do, as it turns out, have physical significance. Next, we may have either of two alternatives. Perhaps there is no physical object at all, and no set of objects, which are associated with this preferred frame (by being, say, at rest in it, or having their motions with respect to it somehow complementary). Or perhaps, as in the Reichenbach–Swinburne worlds, there *is* a set of objects thus associated with the preferred frame, but other observations or experiments show, by Millian methods of eliminative induction, that these objects are *not* causally involved in the cause-effect relationships which pick this out as the one preferred frame. Then we must conclude that what is doing the work is rotations relative simply to this preferred abstract frame of reference itself. Briefly, experiments and observations can show that there are causally significant rotations, but that, as causally significant, they are rotations relative to no *thing* at all. That, I suggest, is what we mean by 'absolute rotation', and the possible observations I have sketched show how we might confirm its reality. It is, of course, a trivial and merely technical task to extend this account from rotations to accelerations of all sorts.

This account shows how, on thoroughly empiricist principles, meaning can be given to such a term as 'absolute rotation'. This is what Locke or Hume, for example, would call a complex idea, and empiricists generally have seen no difficulty in constructing new complex ideas, provided that the materials out of which they are constructed are found within the content of our experience. It is only a 'simple idea' that has to be copied from a preceding 'impression'. Essentially the same principle is expressed in Russell's dictum, "Every proposition which we can understand must be composed wholly of constituents with which we are acquainted".[3] This general empiricism is much more plausible than the verificationist thesis that each form of sentence or statement as a whole must be given meaning by a method by which it could be verified, or, equivalently, that each meaningful term, however complex, must be correlated with some direct observation. That more extreme form of empiricism would preclude the constructing of complex ideas or meanings out of simpler empirical components, and there is no good reason for accepting any such restriction. In the particular case we have been studying, there is a set of possible observations which, interpreted by the ordinary (roughly Millian) canons of causal investigation, would show that it is *rotations* that are at work, and further possible observations which would show that what is causally relevant is rotations relative to no *thing*. These are the constituents out of which the required complex concept of absolute rotation (or, more generally, of absolute acceleration) are put together. (In effect, I am here rejecting what Swinburne, in his reply to Sklar elsewhere in this volume, calls 'verificationism proper', but, like Swinburne, I am accepting what he calls 'word-verificationism', and using it to show how the term 'absolute acceleration' has meaning.)

By analogy with absolute acceleration we can readily explain how meaning can be given to the term 'absolute motion', and how it is at least conceivable that observations and experiments should confirm its reality. If there were some one preferred frame of reference such that even uniform motions (rather than rest) relative to it systematically required causes and had effects, without any *things* associated with that frame being causally relevant to these processes, then we could call motion or rest relative to that one preferred frame absolute motion or rest. To say in general that there is absolute motion would therefore be to say that there is some one such preferred frame; and this might conceivably be confirmed by the discovery of such systematically causally relevant motions without any *things* that could account for their relevance. However, while both 'absolute acceleration' and 'absolute motion' are meaningful, there is dynamic evidence that supports the view that the former term has application, but there is no such dynamic evidence to show that there is absolute motion. A different kind of argument for absolute motion will be put forward in Section 4 below.

## 3. ABSOLUTE DURATION

Another topic with which I want to deal fairly summarily is that of absolute duration. This has not been discussed so widely as that of absolute acceleration, but I have myself written and read (though not published) papers about it on a number of occasions, and while the conclusion I shall reach in this section is not uncontroversial, it my be not too difficult for unprejudiced people to accept.

Let us consider initially how in a single frame of reference – to which our situation on earth and in the neighbouring solar system is a close enough approximation – we decide whether two time intervals *A–B* and *C–D*, marked out by four near-instantaneous events *A, B, C*, and *D*, are equal or not. We use, of course, various sorts of 'clocks', where the term 'clock' can include not only manufactured devices of many kinds but also the rotations and revolutions of planets and satellites, the decay of radioactive materials, and so on. All such clocks are probably inaccurate to some extent, because of causally relevant variations in the circumstances in which they go through their characteristic performances; but we can identify the causes of inconstancy by seeing how similar clocks running side by side may diverge from one another, and once we have found these causes we can either correct the deviant clocks or allow for their inconstancies. For example, we find that of two otherwise similar pendulums swinging side by side, the longer swings more slowly. We can use this knowledge, along with the discovery that pendulum number two is longer during *C–D* than it was in *A–B*, while pendulum number one has remained the same length, to explain why pendulum number two says that *C–D* is shorter than *A–B* while pendulum number one says that these two intervals are equal – that is, swings the same number of times in the two intervals. That is, we can 'correct' pendulum number two on the ground that it, and not pendulum number one, has suffered a causally relevant change. Let us speak of a 'corrected clock' when we are referring to the set of readings given by a clock after allowance has been made for all thus discoverable causes of inconstancy. We can then ask what heppens when we apply to our two intervals *A–B* and *C–D* a large battery of corrected clocks. (The only question we are asking of them is whether they measure these two intervals as equal or not: the obviously arbitrary and trivial choice of units does not come into the matter.) Any one of three things might happen. They might all agree with one another in saying that these two intervals were equal, or, say, that *A–B* was longer than *C–D*; or they might all disagree in a chaotic way; or they might fall into two or more distinct families, with all the corrected clocks of one family agreeing with one another, but systematically disagreeing with the corrected clocks of another family. It is a thoroughly synthetic, empirical, question which of these three results emerges: I can show that the natural procedure for the causal correction of clocks does *not* prejudice this issue by counting as corrected clocks only ones which all agree with one another. Moreover, we can ask what happens when we apply our large battery of corrected clocks not just to a single pair of intervals *A–B* and *C–D*, but to a large number of such pairs of intervals. Now it seems to be an empirical fact that for clocks in this solar system the first of our three possibilities is realised: corrected clocks systematically agree with one another within the limits of observational accuracy in saying, for example, that *A–B* and *C–D* are equal, that of some other pair of intervals the first is greater than the second, and so on.

Now how is this empirical fact to be explained? How does it come about that so many and such different clocks keep on giving agreed answers to a long series of questions of this form? As I said, this result has not been faked up by 'correcting' the clocks in an *ad hoc* way to yield this result. Nor is there any sort of interaction between the various clocks by which they could, so to speak, keep in step with one another. But it is implausible to say that it is a mere coincidence – or rather a whole series of coincidences – that corrected clocks keep on agreeing with one another about the equality or inequality of each pair of intervals. The only plausible explanatory hypothesis is that there is some intrinsic equality between *A–B* and *C–D*, some intrinsic inequality between another pair of intervals, and so on. There are intrinsic metrical relations between such intervals, and that is why each separate clock (once any caused inconstancies have been eliminated) performs the same number of its characteristic performances, whatever they are, in *A–B* as in *C–D*, but systematically different numbers of these performances in other intervals, and so on. This is, in a new sense, the more economical hypothesis.

On these grounds there seems to be a strong case for the absolutist view about the issue numbered 6.2. But we can then go further. Once it is admittend that there are absolute relations of equality or inequality of time intervals, it is an easy further step to the conclusion that these absolute relations are themselves to be explained by the hypothesis that each interval on its own has an intrinsic metrical feature that we can call 'time-length' or 'duration', something that, as we may put it, interacts with the intrinsic features of each clock to determine just what performances it will go through in that interval. In other words, we can regard as well confirmed also the absolutist view on the issue numbered 5.2.

However, this triumph may be premature. This case collapses, this line of argument fails, when we relax our initial restriction and consider 'clocks' which are not all (approximately) in the same inertial frame of reference. When we consider particles, or clocks on space-ships, which are moving, relatively to the solar system, at speeds that are a fair proportion of the speed of light, we have to recognize and allow for the Fitzgerald–Lorentz 'retardation'. The Special Theory of Relativity seems to be empirically well confirmed, and it entails that corrected clocks will conform not to the first but to the third of the possibilities distinguished above. While all clocks which are roughly stationary with respect to the solar system form one family, clocks that are all moving together in a different frame of reference, even another inertial one – say clocks of many different sorts on a rapidly receding space-ship – form another family, and so on. This is dramatically illustrated by the Clock Paradox, for example that the interval which is measured by all corrected earth-related clocks as a hundred years will be measured as only sixty years by clocks on a pair of space-ships which have passed the earth, and one another, as in Figure 1, with four-fifths of the speed of light relatively to the earth. This result is a firm and unavoidable consequence of the Special Theory, and within that theory it has nothing to do with accelerations or decelerations: it depends purely on relative velocities.

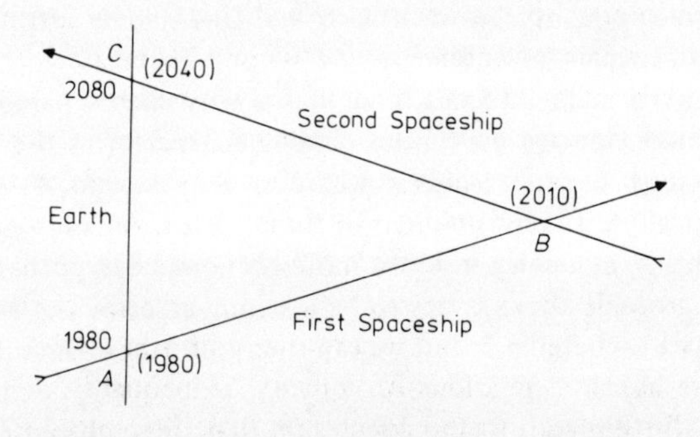


Fig. 1

The Clock Paradox is, however, very easily solved. We have merely to recognize that *clocks do not measure time*. What they measure is something else, 'proper time', or the 'time-like length' of particular space-time paths – the path followed by each particular clock. Whereas our old-fashioned concept is of time as a single dimension, so that there is only one unique time interval between two specified events, such as $A$ and $C$ in Figure 1, the first space-ship's passing earth and the second space-ship's passing earth, once we introduce this new concept of proper time we can see that there can be one proper time interval along the earth's path from $A$ to $C$, but a different – and, as it turns out, shorter – proper time interval along the joint path of the two space-ships $A$–$B$–$C$.

Once we have thus introduced proper time, we can see that the whole of the above argument for absolutism can be transferred to it. What explains the agreements about equality and inequality of intervals among all corrected earth-related clocks? Surely a real, objective, equality or inequality of each pair of slices of proper time along the earth's path. What explains the agreements among themselves of all the corrected clocks of another family, say those on one space-ship? Likewise a real, objective equality or inequality of each pair of slices of proper time along the space-ship's path. And these various equalities and inequalities are themselves explained in turn by the hypothesis that each slice of proper time on its own, each interval along a space-time path or 'world line', has an intrinsic metrical feature of time-like length. That is, we here abandon the absolutist answers to issues 6.2 and 5.2, but accept instead the corresponding absolutist answers to issues 6.3 and 5.3.

It would be only a technical exercise to construct a closely analogous argument with regard to spatial length, first arguing for intrinsic equalities and inequalities of lengths, that is, for the absolutist answer to issue 6.1, then using this to support the absolutist answer to issue 5.1, that there are intrinsic metrical features of length or distance for each spatial stretch on its own, but then abandoning these conclusions in the face of the success of the Special Theory of Relativity, but falling back on another part of the absolutist answer to issues 6.3 and 5.3, that space-like lines in space-time have intrinsic equalities and inequalities, and that each on its own has an intrinsic space-like length. I shall not waste time or space on this technical exercise.

We cannot, of course, appeal to our present *concept* of time as a single universal dimension against the recognition of an indefinite multiplicity of equally real proper times. That concept is exactly what we could have been expected to develop through being acquainted with one and only one proper time, that of things practically stationary with respect to the earth, and through having no experience until very recently of things moving at high speeds in relation to this frame.

The conclusion of this discussion, which I have given in only a condensed form, is that there is indeed an intrinsic metric of space-time. Though the choice of units is of course arbitrary, there are intrinsic absolute quantitative features that have the general character of lengths or distances, though what they are lengths of or distances along are something other than the purely temporal or purely spatial dimensions of Newtonian theory. And it is worth noting that this conclusion rests upon and accommodates and incorporates all the characteristic doctrines of what is called the Special Theory of Relativity. This name is misleading: the doctrines themselves are very far from constituting a pure relativism about space and time.

## 4. ABSOLUTE MOTION AND REST

In this final section I shall put forward an argument that is much more radical than those of Sections 2 and 3. Whereas Sections 2 and 3 were directed only against loose thinking in some philosophical views about scientific matters, Section 4 will challenge what has been an orthodox view for about seventy years within science itself. I shall argue that the Special Theory of Relativity, as ordinarily understood, and as intended by Einstein, destroys itself, and collapses back into the Newtonian picture that includes absolute spatial positions and absolute motion. I have little doubt that my argument in this section is correct. On the other hand, I am not optimistic enough to hope that so heretical a thesis will win widespread acceptance in the foreseeable future. Though, like Hume with respect to religion, I am endeavouring to open the eyes of the public, and look forward to the ultimate downfall of a prevailing system of superstition, I am as reconciled as he was to the reflection that this will not happen in my lifetime, though I need not assume, as he did, that the superstition would last 'these many hundred years'.[4] So I would stress that while the arguments of Sections 2 and 3 may prepare people's minds for that of Section 4, they in no way depend upon the acceptance of the latter. *The conclusions of Sections 2 and 3 will stand, as will the general argument in Section 1 against attempts to settle all the issues together in favour of relativism, even if the more extreme speculations of Section 4 are not cogent.*

I shall restrict my discussion to a type of situation which seems to be adequately represented by the Special Theory of Relativity, not bringing in the complications of the General Theory: that is, situations which are to all intents and purposes free from 'gravitational' forces. My argument starts from the above-mentioned fact, that the Special Theory of Relativity is ill-named, since it is itself far from being a pure space-time relativism. This comes out strikingly in the fact that light paths in space-time, the 'world lines' of light and other electro-magnetic radiations, are physically determinate: *light does not overtake light*. These paths are independent of the source of the radiation. They are equally independent of any observers, and of any choice of frame of reference. If, for simplicity of representation, we neglect two of the three spatial dimensions and picture space-time as a two-dimensional manifold of spatio-temporal points (or possible point-events), possible light paths constitute a rigid grid as in Figure 2. That is, there are two physically determinate families of parallel light paths. (Of course, there are really infinitely many such paths in the two sets of parallels.) This is a very different grid from that presupposed in Newtonian mechanics, shown in Figure 3, where the horizontal lines represent simultaneity at different places, and the vertical lines represent sameness of position at different times. But the Einsteinian grid is no less rigid than the Newtonian one, and it has the advantage of being experimentally detectable, which even Newton did not claim for his. We can tell directly whether two point-events lie on the same light path or not, by seeing whether light emitted at one reaches the other.

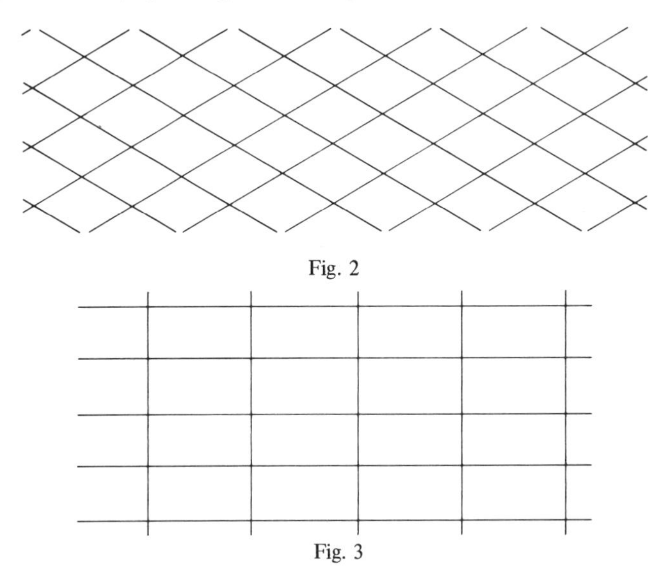

Fig. 2

Fig. 3

Now consider a single point-event 0, where a burst of radiation is sent out in all directions. Again, for simplicity, let us represent merely the light sent out in two diametrically opposite directions, represented as to the left and to the right in Figure 4. An equal group of photons, say, goes in each direction. Then the world lines of these two groups of photons are the determinate light paths in space-time $0A$ and $0B$. Along such a light path there is neither a time-like length nor a space-like length other than zero. If, *per impossibile*, a set of clocks were to travel with the photons, they would all agree in measuring the proper time along any stretch of this path as zero, whether from 0 to $C$, say, or from 0 to $A$. Yet in the going of the photons from 0 through $C$ to $A$, there is undoubtedly a causal process. If the radiation had not been emitted at 0, it would not have been received at $A$. Equally, if something had intercepted it at $C$, it would not have been received at $A$. The photons' being at $C$ is causally intermediate between their being at 0 and their being at $A$. Also, the causal process represented by the line $0A$ is exactly like that represented by the line $0B$ in all respects other than the diametrically opposite directions in which the light travels. It follows that if we arbitrarily select a particular space-time point $C$ on $0A$, there is a unique corresponding space-time point – call it $D$ – on $0B$. That is, there is a part $0D$ of the $0B$ process which is, apart from direction, just like the $0C$ part of the $0A$ process. (See Figure 4.)

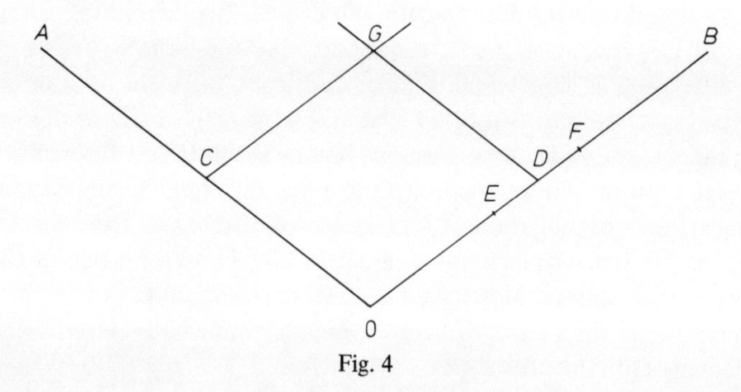


Fig. 4

Here someone might object, asking 'How can you determine this corresponding point?' He might go on: 'Since anything that you could use to measure the "length" of $0C$ would yield the answer "zero", and anything you used to measure the "lengths" of $0E$, $0D$, $0F$, and so on along $0B$ would also measure *each* of these as zero, you could not pick out *which* of $0E$, $0D$, and $0F$, for example, corresponds exactly to $0C$, and hence you could not decide that $D$, say, rather than $E$ or $F$, corresponds to $C$'. I fully concede this. I do not claim that there is any way of *determining* the point that corresponds to $C$; I say only that *there is* one. And nothing but the sort of extreme verificationism which I mentioned, but set aside, in Section 1 would rule out this claim as meaningless. Verificationism apart, there may well *be* things which *we* cannot discover or identify; I persist, therefore, in the assertion that there is, on $0B$, some point that corresponds, in the sense indicated, to $C$.

Now suppose that radiation is sent to the right from $C$ just as the original group of photons reaches $C$, and to the left from $D$ just as the original group of photons reaches $D$. (A mirror at each of these two space-time locations would do the trick.) Then these two new lots of radiation (or reflected photons) will also follow determinate space-time paths, as in Figure 4, and these paths will intersect at a unique, determinate, space-time point, which we can call $G$. The symmetry of the situation shows that, in the single spatial dimension represented in our diagrams, $G$ is symmetrically placed with respect to $0A$ and $0B$. If we were to repeat the procedure, selecting a series of points $C_1, C_2, \ldots, C_n, \ldots$ on $0A$, and identifying the corresponding points $D_1, D_2, \ldots, D_n, \ldots$ on $0B$, and hence deriving a series of intersections $G_1, G_2, \ldots, G_n, \ldots$, then the line $0G_1G_2, \ldots, G_n \ldots$ would be the one and only one world line in this plane which was thus symmetrically related to the physically determinate world lines, the light paths, $0A$ and $0B$. That is, $0G_1G_2, \ldots, G_n \ldots$ represents absolute rest, so far as this one spatial dimension is concerned. Similar constructions yield lines of absolute rest with respect to any two other spatial axes at right angles to this one, and putting them together we have a unique line of absolute rest through 0.

The same point can be made in a slightly different but equivalent way. If a single burst of radiation is sent out in all directions from a space-time point 0, and a space-time point $C$ is arbitrarily chosen on one of the rays, then there is a sphere constituted by the corresponding points on all the other rays – in the sense of 'corresponding' indicated above. Secondary (or reflected) radiation from all over that sphere, coming inwards, will meet at a unique space-time point $G$; then the line $0G$ will be a line of absolute rest.

This conclusion can be related to what was said about the meaning and determination of absolute motion at the end of Section 2. Our present argument shows that *there is* a preferred frame or reference which is the only one to which all the causal processes constituted by the paths taken by radiation (with no 'gravitational' disturbance) are symmetrically related, though there is no *dynamically* preferred frame of reference there is a frame preferred on these other grounds.

This conclusion will, no doubt, seem shocking to anyone with even a slight acquaintance with Relativity Theory. A very natural reaction will be to object that the proposed construction is somehow circular, that the selection of the point $D$ as that corresponding to $C$ must be made from the point of view of some particular observer, or some particular frame of reference, who, or which, has already implicitly taken $0G$ as a line of rest in that frame [5]. But I reply that the light paths $0A$ and $0B$ are observer-independent and frame-independent. They, above all else, are physically real and determinate. I concede that it is only from the point of view of a particular frame or observer that we can *specify* a point on $0B$ which we take to correspond to $C$, a point $D'$ such that $0D'$ *looks* to this observer exactly like $0C$. But I maintain that as a frame-independent and observer-independent fact there must *be* some point $D$ which really corresponds to $C$, which may well not be the $D'$ which from some particular point of view seems to correspond to $C$, and I support this claim by an appeal to the concrete reality of the similar causal processes represented by $0A$ and $0B$.

How can this conclusion be reconciled with the empirical success of the the Special Theory? Very easily. All that success involves is that there is no physical procedure that will pick out a uniquely preferred frame of reference from the set of inertial frames. My argument does not pretend to supply any such physical procedure. It is an abstract, philosophical, argument to show that there really is such a uniquely preferred frame, although we cannot identify it. In a sense my argument makes no difference to physics as a practical concern, as an applied theory. It would not matter at all to someone whose interest in science was purely technological or utilitarian. But it should matter to someone who is interested in the truth, in the question about what is the case.

Though shocking to those familiar with the scientific orthodoxy of the twentieth century, my conclusion should not really be surprising to anyone who reflects upon the way in which, as I said, possible light paths constitute a physically determinate space-time grid. We are simply using the admitted rigidity and symmetry of that grid to show that there must be a further, though concealed, symmetry, in fact to show that there is, though we cannot discover it, the Newtonian grid of Figure 3.

However, there is *something* surprising here. Why, and how, does it come about that although there really is such a thing as absolute rest, we cannot identify it? There must be something odd about the laws of nature that enables them to constitute such a conspiracy of silence. In effect, this means that although there is a unique line of absolute rest through 0 with respect to which the light paths *0A* and *0B* are symmetrically placed, they appear, to each observer, to be symmetrically placed with respect to the line of rest in the inertial frame, whichever it is, with which he is moving. But this is simply the fact that the *measured* veolocity of light in both directions (along *0A* and *0B*) is equal in each inertial frame, when that velocity is measured by the means which it is natural for an observer at rest in that frame to use. In effect, we shall have to take the Fitzgerald–Lorentz 'contractions' and 'retardations' literally, as real effects of high-speed absolute movement, that is, as Fitzgerald and Lorentz themselves interpreted them. But this is only a hint; a rather long story would be needed to explain in detail how mere motion could have such effects.[6] And I agree that until such an explanation is given, this conspiracy of silence will provide the best available evidence for the existence and activity of Descartes's *malin genie* – though I would, for this purpose, translate '*malin*' not as 'malignant' but merely as 'mischievous'.

When we thus take the 'contractions' and 'retardations' literally, as effects of absolute motion, we are explicitly satisfying the definition of 'absolute motion' that was offered at the end of Section 2, at least to the extent that we are saying that there is a preferred frame of reference such that even uniform motions relative to it have systematic effects, quite independently of there being any things associated with that frame.

If the argument of this section is sound, and there is such a thing as absolute rest (and therefore also absolute spatial position), this not only settles the issues numbered 3 and 2.1 in favour of absolutism, it also reacts upon the decisions we reached at the end of Section 3. At that point I said that the belief in absolute durations – lengths of time rather than proper time – could be rejected as resulting from the fact that our experience, until recently, has been confined to the proper time of one particular inertial frame, which we have therefore mistaken for a single universally applicable dimension. But it now appears that out of all the different proper times there will be a unique one with a special status, namely the proper time of things at absolute rest. This we can now call 'time' *simpliciter*, and we can call the determinate lengths of this time 'absolute durations'. Other proper time intervals will still be intrinsic features of the space-time paths, but as resultants of their spatial and temporal components, and their significance as measurements will be a consequence of the odd behaviour of clocks systematically affected by Fitzgerald–Lorentz retardation. These clocks can therefore be corrected (as in the early part of Section 3) for these causal disturbances to yield measurements of *time*, and then *all* corrected clocks will, after all, come into line with one another. However, this unification of all families of clocks is purely theoretical: we can make no practical use of absolute time. In particular, we have no right to assume (at least without some further argument) that the proper time of our solar system is time *simpliciter*. Though our old-fashioned belief in a unique time dimension will turn out to have been true, it will still have been, as I said at the end of Section 3, unjustified.

To conclude, then, there is a case for absolutisms about each of the issues numbered 2.1, 3, 5.1, 5.2, 5.3, 6.1, 6.2, and 6.3. About that numbered 4, my specific conclusion in Section 2 was only that it is an empirical question whether there is absolute acceleration or not, but if Section 4 shows that there is absolute motion, this will carry absolute acceleration with it. I have said nothing specifically about issue 2.2, or about any of the issues concerning absolute or relative existence.

However, I am at least as much concerned about the methods used in this discussion as about the conclusions reached on particular issues. My general thesis is that items that are not directly observable, and terms and statements to which meaning cannot be given directly by methods of verification, can be legitimately introduced. To justify such introductions, we require only familiar methods for the sorting out of causally relevant factors, together with the principle that a hypothesis which would compel us to accept certain observed agreements as no more than a massive coincidence can be rejected in favour of a hypothesis which explains that apparent coincidence by resolving the agreements into consequences of some unitary state of affairs.

## NOTES

[1] See, for example, my 'Truth and Knowability', *Analysis* **40** (1980), 90–92.

[2] I. Newton, *Principia*, Scholium following Definition viii; E. Mach, *The Science of Mechanics*, LaSalle, Illinois, 1960, Chapter 2; H. Reichenbach, *The Philosophy of Space and Time*, New York, 1957, Chapter III, § 34; R. Swinburne, *Space and Time*, London, 1981, Chapter 3.

[3] B. Russell, *The Problems of Philosophy*, London, 1912, Chapter 5.

[4] Letter from Adam Smith to William Strahan dated November 9, 1776, in *Hume's Dialogues concerning Natural Religion*, ed. by Norman Kemp Smith, London, 1947, pp. 243–8.

[5] [Mr. Mackie included at this point a sentence stating that the above-mentioned objection would be discussed in the additional note which he proposed to write at the end of the article. – *Ed.*]

[6] I am indebted for some suggestions about how such a story might go to the late Geoffrey Bulder, of the Sydney University Physics Department.

JON DORLING

# REPLY TO MACKIE

## A. OUTLINE OF MY OWN PHILOSOPHY ON POSITIVIST-REALIST ISSUES IN PHYSICS

I believe that God[1] is an extreme positivist, and presently I shall prove it. I also believe that we should try to emulate His system of beliefs. It follows that we should go in for extreme positivist critiques of existing theories in physics, and that this will yield a real prospect of improving them. But it also follows that we should not adopt an anthropocentric form[2] of positivism. We need not, then, have positivistic doubts about the reality of the past, of physical objects, or of the microworld. I also happen to believe that the final correct theory of the universe will be almost unbelievably simple when axiomatized in an extreme positivist form.

A scientific theory involving certain theoretical entities may stand in four possible relations to a more positivist version of the same theory:

(1) Neither the existence nor the uniqueness of the disputed theoretical entities is provable within the more positivist axiomatization.

(2) The existence, but not the uniqueness, of the disputed theoretical entities is provable within the more positivist axiomatization.

(3) The uniqueness, but not the existence, of the disputed theoretical entities is provable within the more positivist axiomatization.

(4) Both the existence and the uniqueness of the disputed theoretical entities is provable within the more positivist axiomatization.

Surprisingly, I know of examples of case 1 only outside physics, and the only example I know of case 3 was Bertrand Russell's celebrated positive contribution to theology[3]. Case 4 is, contrary to what philosophers might expect, very common in physics:

Thus the existence and uniqueness of the values of the functions specifying forces and masses, is in fact *provable*, for our actual world, within most positivistic axiomatizations of Newtonian mechanics.

Again, it is *provable* within Poincaré's extreme positivist relative–accelerations-only axiomatization of Newtonian gravitational theory, that there exists the usual class of frames, unique up to a Galilean transformation, with respect to which Newton's laws are recovered in their simple, less positivistic, form.



*R. Swinburne (ed.), Space, Time and Causality*, 23–35.

Furthermore, cases of type 2, and these include the cases of potentials in electromagnetism, of fields in electromagnetism, of the Lorentzian aether, and of Newton's own version of absolute space, can be reduced to cases of type 4, provided the realist party to the dispute adopts appropriate boundary conditions, or otherwise guarantees uniqueness, e.g., by tying the aether frame to the inertial frame of the universe as a whole.

When I say that the realist's existence claims are provable by the positivists in cases 2 and 4, I am talking about ordinary mathematical existence theorems; the realists however seem to want something more, and consideration of case 4 makes it clear what this is.

This issue in case 4 is whether we should allow what, from the positivist's point of view, are complex defined terms, in our axioms. Well, our choice on this issue determines how we will order alternative possible worlds in terms of plausibility and simplicity. Hence it determines, inter alia, (i) which possible world we take as the simplest consistent with the existing data, and, more importantly, (ii) what we should regard as the next simplest, next simplest but one, and so on, possible world, should our present theory prove observationally incorrect. In other words, it determines, inter alia, what further experiments we think would be most worth doing, which would constitute the most severe tests of the theory, where we think the theory is most likely to break down. So even though the observational contents of the realist version and the more positivist version of the theory are exactly the same, the realist and the positivist, so long as they do not think the existing theory is certainly the whole truth and nothing but the truth, differ, albeit slightly, in their observational expectations. And in this strict sense their dispute is really an empirical dispute.

Now I say, and this is the crucial point, that if someone believed – as no scientist ever has to date – that the theory, he had, contained the whole observational truth and, observationally, nothing but the truth, then he would have no choice but to adopt the extreme positivist version of it. For given a type 4 choice, there could be no real gain, in simplicity of world description, by adopting complicated defined terms in the axioms. For one can always make a theory as simple as one likes by using complicated enough defined terms. Hence the realist option here is merely a re-writing of the positivist one. (What such a theorist hopes in our actual scientific situation is that such complex defined terms will prove more simply definable in a more correct or more complete theory, but by hypothesis this option isn't open to him in the hypothetical situation presently under discussion.) Secondly, given a type 2 choice, the uniqueness claim required by the realist alternative is simply false, in the imagined hypothetical situation, *qua* mathematical uniqueness claim. A function would be claimed to be unique which was provably not unique. In the normal situation the uniqueness claim could be true in virtue of facts the theory left out, but not in the idealised situation of the theory supposedly containing the whole observational truth and observationally nothing but the truth. (Mackie's aether theorist would here be reduced to asserting the uniqueness of a frame which was provably non-unique within his own theory. And this impasse could not be blocked by any attempted distinction between mathematically unique and physically unique, or between mathematically preferred and physically preferred, once the theory was supposed to contain the whole of physics.)

So I not only maintain that God is an extreme positivist in physics, I have just given a proof of it.

However, in our actual scientific plight, we do not believe that any existing theory is observationally the whole truth and nothing but the truth, so the uniqueness claim, accepted by the realist and rejected by the positivist in case 2, may actually be true and provable in God's theory. And in case 4, it may be that, given an observationally more complete and more correct theory, what is complicated to define in our existing observational theory, will prove very simple to define in the correct theory, and that actually it is the positivist's choice of primitives, which will turn out to be the more complicated defined terms, when the final observational truth is known. So to be a realist over a given entity is, if you are rational, precisely to believe that the observational content of the existing theory is almost certainly incorrect or incomplete *in a certain direction*.

Now I say this is exactly what happened, through subsequent developments, with the key theoretical entities in Newtonian mechanics. Absolute accelerations, being only dynamically definable, were enormously complex defined terms, in Newton's only legitimate observational language (i.e., in the one determined by the theory). But a small later correction of his kinematics, to give special relativistic kinematics, turned absolute acceleration into a very simply defined, purely kinematic, term. The same story is true, less dramatically for measures of temporal and spatial congruence, in the transition from Newton to SR.

Something similar is almost certainly true of masses, which can really just be interpreted as rest-frame frequencies in relativistic quantum mechanics.

So I, in spite of being an extreme positivist, say Newton was actually right in introducing all these theoretical terms in his axioms. He was also right in thinking that absolute accelerations and durations should be thought of as essentially kinematical entities, even though, within his theory, he could only define them dynamically.

As a result of Newton's realist choices here, I say we can still accept all three of his laws of motion (reading rest-masses, proper accelerations, and so on) within special relativity, and even (reading geodesic, as well) within general relativity: we need only to change Newton's kinematics, not his dynamics. Newton's positivist critics were wrong, and their suggested directions of change would have led us further away from God's true theory. Had they understood the logic of positivist-realist disputes more clearly, they would not merely have recognized, as they did, that Newton's theory, like all realist theories in physics, contains the seeds of its own destruction, but they might have spotted earlier, the arbitrariness of Newton's kinematic assumptions, changing which, so as to bring space-time geometry more closely into line with ordinary spatial geometry, enabled Newton's realist intuitions concerning the fundamental rightness of his foundation of dynamics, to be later confirmed by a truer, more positivist, theory.

A good positivist is a realist who recognizes that, in so far as his theory is realist, it is false.

## B. COMMENTS ON MACKIE

### 1. *Special Relativity and Absolute Motion*

First I regard the thesis which John Mackie defends here, as prima facie unlikely, on quite general philosophical grounds. I can believe that a particular formulation of a physical theory, even 'the orthodox formulation', may contain unsuspected metaphysical elements, in the sense that it commits us logically to assigning definite truth values to one or more statements, for which it makes no experimental difference which truth value is assigned. Of course what usually turns out to have been 'unsuspected' in such cases is not the presence of the elements in question, but their immunity from experimental evidence. Or it might have been accepted that the orthodox version of the theory was obviously prima facie committed to such elements, but have been *wrongly* supposed that it was easy in principle to eliminate them without any drastic change in the structure of the theory (consider for example the indeterminacy of the phase of a wave function and the apparent determinacy of the vector potentials in orthodox quantum mechanics). But Mackie seems to be claiming much more than this. He seems to be claiming that although the axioms of orthodox special relativity are invariant under a certain group of transformations, they have consequences which are not invariant under those transformations. That seems to me impossible if Mackie's 'philosophical' consequences really are logical consequences. Furthermore Mackie seems to be making his claims not just about one particular formulation of orthodox special relativity theory, but about any formulation of it whatever, which preserves its empirical, experimentally checkable, content. But that, again, seems to involve us in contradiction.

Mackie considers light sent out in two diametrically opposite directions from a point event 0. (As a matter of fact, two different directions would be enough for Mackie's argument; there will always be an inertial frame, indeed an infinite class of distinct inertial frames, which will make those directions diametrically opposite.) He now argues that there are like causal processes 'represented by' the two null lines $0A$ and $0B$, and hence that if we select a point $C$ on $0A$, there must be a point $D$ on $0B$, such that the part $0C$ of the $0A$ process is just like the part $0D$ of the $0B$ process. And he insists that the point $D$ must be unique. I agree that, except in a rather hypothetical case which I can just imagine, it is unique. Though I deny its uniqueness follows from Mackie's assumptions. It is unique because light is not structureless. We could count wave crests along the null lines, and choose points of like phase (and direction of change of phase). This uniqueness could only breakdown if, say, we superposed enough different contributions of different frequencies or phases to make the crests disappear altogether and to make the light essentially structureless. (This was the rather hypothetical case I could just imagine.) But let us take for granted the uniqueness. What does it give us. It allows us, as Mackie correctly argues, to define a unique inertial frame associated with that burst of radiation. But contrary to Mackie, this doesn't show that there is a preferred frame of reference, it merely shows that there is a preferred frame of reference associated with that burst of radiation. It is in fact merely the, empirically easily determined, frame in which the light going in both directions has the same colour. In any other frame the light going in one direction will be Doppler shifted to the red, and the light going in the other direction will be Doppler shifted to the blue. But it is not an absolute frame, because it will be different for different bursts of radiation. It is in fact the centre-of-mass frame of the system of particles which emitted the radiation. (If other radiation is emitted at the same time, the post-emission centre-of-mass frame may of course be a differet frame.)

I conclude that Mackie's attempt to convict orthodox special relativity theory of a hidden commitment to an absolute rest frame, hence to absolute motion, doesn't work, and I have already suggested that there are general reasons why no such argument could work.

## 2. *Absolute Intervals*

There are two things that I find peculiar about Mackie's defence of absolute intervals. First is his route from corrected clocks to absolute equalities and inequalities of intervals. Second is his route from the latter to absolute intervals.

I am not clear how Mackie corrects his clocks. He says "all such clocks are probably inaccurate to some extent . . . but we can identify the causes of inconstancy by seeing how similar clocks running side by side may diverge from each other, and once we have found these causes we can either correct the deviant clocks or allow for their inconstancies". I know how I correct clocks. I use quantum field theory, or if the radiation field is irrelevant, the Dirac equation, and if the velocities are small enough the Schrödinger equation, and if the action is large compared with Planck's constant, Newtonian mechanics or general relativity theory. I don't know how to correct clocks unless I can use a theory which contains a time parameter, or various time parameters. But Mackie cannot do this, because it would make his whole argument circular. How then does he do it? Suppose he has two pendulums, a long and a short one, and they fail to keep time with each other. What can he do? Perhaps he intends by similar clocks, ones which are prima facie identical, but which may nevertheless fail to keep time with one another. But I still don't understand this business of identifying the causes of inconstancy, without any theory containing time as a parameter.

Even if we grant Mackie clocks corrected in his sense, it seems to me he will be letting in families of clocks corrected in *his* sense, which are not families of clocks corrected in *our* sense, and I cannot see the slightest reason why different families of his clocks should agree with each other on equalities and inequalities of intervals. Surely his prescription must let in families of clocks which, for perfectly good physical reasons, all progressively go slower and slower, in time with each other.

It seems to me that the alleged empirical fact that Mackie is trying to conjure up here is better expressed by saying that all processes in nature can be described either by a single physical theory containing a time parameter, or by a collection of physical theories containing the *same* time parameter. Now I don't see that that is a fact which *has* to be *explained* by postulating absolute equalities and inequalities of intervals. One *could* do the latter by taking space-time geometry as prior to and as presupposed by dynamics, but one could equally adopt Minkowski's own view that dynamics is more fundamental and is self-sufficient: space-time geometry is then just a picturesque way of describing some of the transformation properties of our laws of dynamics. I prefer to treat the space-time geometry as having an explanatory role, but I am very aware that I may be making a bad mistake, and I don't know any very persuasive argument for my position. I don't see clearly what Mackie's postulation of absolute equalities and inequalities of intervals comes to, unless it is a matter of siding with me against Minkowski on this issue. (For those on whom the historical allusion to Minkowski is lost, I should say that I read him as having wanted to reduce geometry to physics, not vice-versa: that is one reason why he had no interest in presenting his geometry as a synthetic (i.e., axiomatic) geometry.)

What about Mackie's step from absolute equalities and inequalities of intervals to absolute intervals? The odd thing about this is that Mackie turns it into a peucliar philosophical inference when in fact, just as in Euclidean geometry, you can *define* absolute intervals (modulo a choice of unit interval) in Minkowski geometry, once you have absolute equalities and inequalities of intervals. But Mackie says that once you have the latter, "it is an easy further step to the conclusion that these absolute relations are themselves to be explained by the hypothesis that each interval on its own has an intrinsic metrical feature that we can call its 'time-length' or 'duration' ". So Mackie thinks the step to absolute intervals is somewhat of the form $p \rightarrow q$, $q$, therefore $p$, very plausibly. Now I don't suppose Mackie would wish to defend such an inference form in general, but there is a suspicion throughout his paper that, once one has rejected the dreaded positivism, $p$ here is just a metaphysical assumption which costs you nothing and gives you an explanatory gain. There is the suspicion that Mackie would be equally willing to defend the move from absolute acceleration to absolute velocity to absolute position for the sake of a similar explanatory gain.

Let us suppose though, that Mackie knows that absolute intervals can be defined (uniquely modulo the choice of unit) once you have the absolute equality/inequality relation between intervals. Then there is still the question of whether his is the right direction of explanation. It is not Euclid's direction of explanation: Euclid wished to explain the quantitative in terms of the qualitative. Now I happen to think that the right move here is to accept the need for *ultimately* explaining the quantitative in terms of the qualitative, but to reject Euclid's particular qualitative explanation on the grounds that there are valid objections to taking the notion of a straight line as a primitive notion and equally to taking any of its surrogates, such as betweenness, as a primitive (in the present context the analogous objections come from the fact that classical free particle motion, needed to interpret straight time-like line if that is to be a primitive, is only a classical approximation to reality: at a more fundamental quantum-mechanical level, the metric seems more important than the geodesics). But Mackie doesn't have any special arguments of this sort. He hardly wants to argue that he wishes to retreat philosophically from the purely qualitative to the quantitative, in order, hopefully, later to advance again to the qualitative in another direction, and hence later recover the philosophical ground lost. So it doesn't seem to me that *he* has any good reason for choosing that particular direction of explanation.

It might seem that the choice of direction of explanation here is just a choice between different inter-definable sets of primitives and that *hence* it cannot really matter to physicists and is just a sterile philosophical dispute. This is not so. In fact if you adopt a Euclidean style axiomatization of special relativistic space-time geometry, then *even* with a strong axiom of continuity, you need an *independent* axiom of Archimedes. If you then take the view that this is a bit of an excrescence, since it is only needed to prove that null lines really do have a totally degenerate congruence structure, and you drop the axiom of Archimedes to see what else is then let in, it turns out to be the possibility of additional non-temporal non-spatial dimensions with a discrete structure but geometrically tied up with space-time (mass? charge? – I have never bothered to work out the predictive content). So in fact one finds there is a now a natural way of simplifying the axioms with a potentially novel predictive content. But had one chosen instead a primitive metrical notion for ones axiomatization, such predictions could not have arisen in any natural way, i.e., there would be no simple modification of the axiom system which yielded them and left nearly everything else intact, so they would have seemed most implausible. Quite generally the choice of primitives for the axiomatization of a physical theory alters the simplicity rank-ordering for alternative possible worlds to the one described by the theory, and hence determines what the theorist tells experimenters to go out and search for.

### 3. *Verificationism and Absolute Acceleration*

Until very recently, I would have given a very similar account of absolute acceleration in Newtonian mechanics to John Mackie's account, but I would have described what I was doing as explaining the meaning of absolute accelerations by describing their method of verification. It seems to me that that is just what John Mackie is doing. (I do not understand his putative criticism of verificationism on this issue; he gives a characterization of verificationism in terms of 'each *form* of sentence' or 'equivalently' in terms of 'each meaningful term' which I neither recognize as verificationism nor understand. I thought verificationism simply said that the meaning of any sentence or statement was given by its method of verification, and said nothing about forms abstracted from sentences or terms extracted from them.

Now I want to insist, contrary to Mackie, that absolute acceleration does not exist, it is meaningless, if, and this is an important proviso, Newtonian physics, including Newtonian kinematics, is true. Perhaps it is better to say, absolute acceleration would not exist, would be meaningless, were Newtonian kinematics true. (As a matter of fact Newtonian kinematics is false, and absolute acceleration does exist, and is perfectly meaningful, if special relativistic kinematics, or even general relativistic kinematics, is true.) Compare the case of absolute slope, i.e. the slope of a line, but not relative to any other line, or to any plane. Absolute slope does not exist, it is meaningless, if Euclidean geometry is true. On the other hand absolute curvature does exist if Euclidean geometry is true: we can, within Euclidean geometry, define the curvature of a line, and not just the relative curvature relative to some other line taken as standard of straightness.

However something else, which I would be prepared to call inertial acceleration, or dynamical acceleration, does exist, is perfectly meaningful, if Newtonian dynamics is true; and Newtonian dynamics is true. (Special relativity does not alter Newton's dynamical laws, once we understand them against a background of the correct kinematics; even general relativity only alters the inverse square law, not Newton's laws of motion. Newton's first law does not refer to straight lines (in the Latin) but to infinitesimal parallel displacements (uniformiter in directum, not secundum lineam rectam).) I refuse to identify intertial (or dynamical) acceleration, with absolute acceleration, because absolute means either perfect, or not relative, and inertial acceleration is neither perfect nor not relative.

Inertial acceleration is a primitive term in conventional formulations of Newtonian dynamics, but it can be explicitly defined as a logical construction from directly observational primitives, within an extreme positivistic formulation of Newtonian kinematics and dynamics. The latter is so complicated (the force-laws contain second and third order derivatives of the relative positions of bodies) that it is virtually impossible to believe that it can be the ultimate truth about reality. One way out is to suppose that the laws of Newtonian *dynamics* are incorrect or incomplete: this was the relativist programme of Mach and later of Einstein. Another way out, the correct way out, was to put the blame on the kinematics, and to suppose that existing kinematics was incorrect or incomplete. This was Newton's choice, namely to treat absolute accelerations as a new primitive or new defined term in kinematics. Newton took it as a term defined with the help of the new primitive absolute position, because he was convinced by a fallacious argument leading from absolute accelerations to the necessity of absolute velocities, and hence to the necessity of absolute positions. He then tried to extend the observational language of kinematics to include absolute positions, by introducing God as a sort of super-observer with observational capacities going beyond mere human ones. But that is a cheap philosophical trick, since one cannot show that God can observe a distinction which has not been shown to exist. God would not have helped, even if Newton had more legitimately restricted himself to introducing absolute accelerations themselves as a new kinematic primitive. Newton should instead then have inferred that the empirical content of existing kinematics was *either* false, *or* radically incomplete in a way that the mere introduction of God as a super-observer could not remedy, but which required the empirical discovery of new observables additional to ordinary spatial and temporal magnitudes. In fact it was false.

Were Mackie right in his view that correction of clocks does not presuppose dynamics, then the situation would now have been simple and clear. Unfortunately, Newton rightly recognized that temporal congruence, and perhaps also spatial congruence, is not well-defined without appeal to dynamics. Newton could then argue that just as we cannot observe whether one clock is speeding up or slowing down *relative* to another, unless we appeal to dynamics to determine the standard of equal intervals of time, to decide which is really doing what; so we can only observe relative accelerations, unless we appeal to dynamics to determine what is really accelerated. Thus he could argue that the kinematic notion of absolute acceleration presupposed dynamics in no different a way from the kinematic notion of equal time intervals, which was certainly not definable in terms of genuine purely kinematic observational primitives.

However, if my approach is right, this is just a second source of trouble in Newtonian physics. We are entitled to talk about dynamically equal time intervals, and this is a complex defined term in Newtonian dynamics, but the dynamical theory then needs to be reaxiomatized in terms of its genuinely observational primitives. If the theory then appears unbelievably complicated, we must suppose that it is either empirically false or radically incomplete. We can then either put the blame on the dynamics, or on the kinematics. Newton again made the right choice of putting the blame on the kinematics, but he again opted for radical incompletences, and tried to repair it with the help of God. In fact, of course Newtonian kinematics was simply false, and the correct kinematics (for our present purposes, special relativistic kinematics) does enable one to define both absolute accelerations and temporal congruence in terms of genuinely observational kinematical primitives. Only now are we genuinely entitled to talk about (absolute) accelerations, or accelerations, simpliciter, and equal time intervals, rather than inertial, or dynamical, accelerations, and dynamically equal time intervals: we no longer have to rely on a pious, and not clearly warranted, because there were alternative ways out in terms of a revision of dynamics, belief in the falsity or radical incompleteness of conventional kinematics. Absolute accelerations are just the curvatures of world lines: these curvatures were not definable within Newtonian time-geometry: they are straight-forwardly definable within special relativistic time-geometry. Equal temporal intervals were not a genuinely observational primitive, nor definable in terms of a genuinely observational primitive, within Newtonian kinematics: but temporal and spatial congruence can both be defined on the basis of a single primitive observational concept of temporal succession, within special-relativistic space-time geometry. (This was noticed first by A.A.Robb around 1911, and republished by him many times in the next twenty-five years, and then forgotten until E.C. Zeeman claimed and largely obtained the credit for it fifty years later.)

This is not of course the end of the story. A further positivist critique of our existing theories is certainly possible. All existing theories which become unbelievably complicated when expressed in terms of the simplest observational primitives, contain important clues to their own ultimate falsity or incompleteness. Very few of Einstein's specific criticisms of the theories he destroyed were wholly correct, but he was right in thinking that an empirically apparently adequate theory can contain the seeds of its own destruction, and that these can be revealed by a conceptual critique.

## 4. *Absolute Locations and Absolute Moments*

John Mackie apparently believes[4] that the universe could have come into existence 24 hours later than it did and in a position a mile to the north of its actual position. (So as to avoid footling objections about 'north', I shall take it he means everything translated a mile in a direction parallel to the earth's present axis.)

Now judging from his later discussion of absolute acceleration, the arguments he would offer for this would at best establish that this *would* make sense if the correct dynmaical theory were neither invariant under translations in time, not invariant under spatial translations and rotations. But of course that doesn't show that it *does* make sense, it merely shows under what conditions it *would* make sense. If Mackie is to *believe* that his is a real possibility, he must not merely believe that such a dynamical theory is a real possibility, he must believe that such a dynamical theory is actually true, hence that the theories that physicists and astronomers presently believe in are actually false, and in a very specific way. Possibly possible does not imply possible in this context.

Secondly, if my earlier considerations were correct, he ought then merely to speak of 'dynamically 24 hours later', 'dynamically a mile to the north'. Unless of course, he is prepared to go further and believe in a revised kinematics which is not time-translation invariant and in a revised spatial geometry which is not invariant under translations and rotations. But then he must either believe that our present kinematic theories are actually false in a rather specific way, or he must believe that there are, yet to be discovered, additional observational primitives which will destroy the present invariance properties of temporal and spatial geometry.

## 5. *The Principle of Sufficient Reason*

Mackie's criticism of the principle of sufficient reason is very brief. At first reading it seems as if perhaps he is not criticising the principle itself, but merely the version which brings God in, and says that God does nothing without a sufficient reason. But on careful reading it is clear that it would make no difference to Mackie's case, if God were left out of the argument, and all that were maintained was that nothing happens, or is the case, without a sufficient reason.

Now Mackie's only criticism of this principle is that there is no sufficient reason why we should accept it. If he means, by this, no logically compelling reason, then of course he is right. But it seems to me that that is hardly to the point if we nevertheless have good reasons for believing the principle to be true.

Suppose the advance of science constantly shows us that things which at first appeared to happen, or to be the case, for no sufficient reason, later turn out to happen, or to be the case, for some good reason or other. Then it seems to me intelligent to guess that this may be quite generally the situation, and we can then use this principle to criticise existing theories. Indeed I think it is constantly used in subtle ways by theoretical physicists in their striving for ever better theories. Taken to its logical conclusion, it does of course entail that our existing universe will ultimately turn out to be the simplest possible universe. But I see nothing wrong with that.

## NOTES

[1] All references to God in this text are a mere facon de parler and eliminable.
[2] How can I distinguish theoretical from observational terms without antropocentrism? I take a theory itself to provide the conditions for determining which of its terms are observational, or *rather* to provide *sufficient* conditions for certain terms not being observational. These terms, as positivists, we must eliminate. As we later extend our theories, to more complete theories of the universe, additional terms may turn out also to have a non-observational character, in spite of appearances, and also have to be eliminated. I don't justify this conception here, but it seems to fit my examples.
[3] Russell proved from logic that there was at most one God, but promised a later proof that there was not even one.
[4] He has subsequently denied that this view should be attributed to him, though it still seems to me that it is defended at more than one point in his text.

ELIE ZAHAR

# ABSOLUTENESS AND CONSPIRACY

I find myself uncomfortably holding a position which is halfway between Mackie's and Dorling's. I am in full sympathy with Mackie's view that absolutism cannot be ruled out on general verificationist grounds. Anyway, I thought – rather naively – that operationalism and strict verificationism were long dead. The universal quantifiers, which most respectable theories involve, constitute to my mind an insuperable obstacle to verificationism; unless of course it is assumed that the domain of discourse of such theories in finite, or at least discrete; an assumption which is unverifiable. As for operationalism, I took it for granted that the attempts to reduce *all* theoretical concepts to empirically decidable ones had failed; unless one weakens the operationalist requirement by admitting entities which are *indirectly* detectable; but then all concepts occurring in an empirically testable theory can be regarded as being 'observational' in this wider sense of the word. The operationalist criterion would break down for it would then fail to demarcate. This is why I cannot agree with Jon Dorling when he says:

> Now judging from his later discussion of absolute acceleration, the arguments [Mackie] would offer for this would at best establish that this *would* make sense if the correct dynamical theory were neither invariant under translations in time, nor invariant under spatial translations and rotations.

This passage is clearly question-begging. Only if one adopts a verificationist principle of meaning does Mackie's argument cease to make sense. According to the correspondence theory, truth, and in particular ascertainable truth, presuppose meaning, not the other way around. There is nothing meaningless or nonsensical about Mackie's claims concerning absolute rest, even if these claims were empirically untestable. I happen to believe that such claims are false (hence meaningful) because they clash with *fruitful* metaphysical principles which have guided *empirically* successful programmes.

Jon Dorling puts forward a very interesting, though to my mind highly suspect, conjecture. Take two scientific hypotheses $T$ and $T'$; reduce them to two other theories, $T_0$ and $T'_0$, which contain only observational predicates. Thus $T$ and $T'$ are observationally equivalent to the two positivistic theories $T_0$ and $T'_0$ respectively. Should $T_0$ turn out to be more complicated



*R. Swinburne (ed.), Space, Time and Causality*, 37–41.

than $T'_0$, then we are invited by Jon Dorling to conclude that $T$ is more likely to be false or incomplete than its rival $T'$. I have already indicated why, in the most interesting scientific cases, the reduction of $T$ and $T'$ to $T_0$ and $T_0'$ is impossible or at any rate highly problematic. Even if such a reduction were possible, I fail to see what reasons Jon Dorling can adduce in support of a thesis which is not purely philosophical but empirical. After all, $T$ and $T'$ can be tested; and it may turn out, despite the relative simplicity of $T'_0$, that $T$ and $T_0$ are confirmed while $T'$ and $T'_0$ are refuted. It may moreover be the case that, though $T'_0$ is simpler than $T_0$, of the *four* hypotheses: $T$, $T_0$, $T'$ and $T'_0$, $T$ is the simplest one.

I suspect that Jon Dorling's claim stems from the positivist prejudice both that reality is coherent and that it wholly consists of directly ascertainable facts; hence the possibility of describing it by means of simple theories restricted to the observational level. Let me confess to a different prejudice. From an evolutionary point of view it can be argued that only a small portion of reality is directly observable, namely that part which is urgently needed for survival. To think otherwise is to assume that everything is relevant to our survival, that the universe is there for our sake, which seems both untenable and somewhat megalomaniac. Thus, even if reality formed a coherent whole, it is highly unlikely that the fragments of it which impinge on our senses should by themselves fall into a simple pattern. Only by going beyong what is immediately given could we hope to integrate the whole picture.

Let me now turn to what I regard as Jon Dorling's justified criticisms of Mackie's position. Dorling's remarks about a number of particles determining their own inertial system and *not* an absolute rest frame seem to me ungainsayable. He also rightly points out that Mackie dismisses the principle of sufficient reason a little too lightly. That this principle constitutes no logical truth is a point in its favour, which shows in effect that it has bite, that it is non-vacuous. The fact that the principle of sufficient reason does not apply to itself hardly constitutes a valid criticism of it, or of any other postulate for that matter. I fully agree with Dorling that certain (in my sense metaphysical) axioms can guide or constrain research programmes so as to increase their empirical content. It can be argued that the principle of sufficient reason, in the form of a covariance requirement, was and still is a powerful heuristic tool for the construction of physical theories. It can also be argued that the metaphysical postulates which govern empirically successful hypotheses are indirectly supported by the success of such hypotheses. Mackie concedes that his argument about absolute position and time "make no difference to physics as a practical concern, as an applied theory".

It seems to me that the situation is in fact much worse than he admits: since covariance applies at the most fundamental level, Mackie's conclusions not only do not affect *theoretical* as well as applied physics, but are also at odds with the spirit of the whole relativistic enterprise. His views are therefore undermined to the same extent that Relativity is confirmed by the facts. (By Relativity I mean covariance.) Mackie faces the same puzzling problem which confronted Lorentz. If one postulates, or philosophically defines, an absolute frame of reference, then one has simultaneously to accept, in Mackie's own words, a huge "conspiracy of silence". Nature conspires systematically to conceal from us the asymmetry which marks off one privileged frame from all other inertial systems. This explains why, until 1909, Lorentz hoped that Maxwell's equations would turn out not to be fully covariant; once he realized that they were, and that the set of Lorentz transformations formed a group, he conceded defeat [1]. He still hankered after the ether, but he honestly accepted covariance as the *operative* principle. Since the inverse of a Lorentz transformation is itself a Lorentz transformation, every inertial observer could, with total impunity, regard himself as being at rest in the ether. To accept an absolute rest frame was to accept, in Cartesian terminology, that God was a systematic deceiver. In Einstein's jargon, God would not only be very sophisticated, he would also be malicious. This argument can however be recast without any reference to God; Einstein was after all no ordinary theist, he did not believe in a transcendent Supreme Being. One could argue as follows: it is unlikely that Nature contains both deep asymmetries and compensatory factors which exactly nullify these asymmetries. Such a state of affairs is not logically impossible and to envisage it is not meaningless; but it is unlikely, or improbable, in the same intuitive sense in which a series of coincidences and accidents having a single global effect are improbable. This sort of metaphysical argument, or principle of sufficient reason, is successfully used both in the sciences and in everyday life. Would one not feel sceptical if one were told of a certain man that he was an intrinstically good person, but that, every time he embarked on any course of action, some hidden psychological mechanism impelled him to do the morally wrong thing? Would one not be tempted so to redefine goodness as to be able to call him an immoral person? This is precisely the attraction of the positivist solution, which unfortunately overkills the problem. This is also why Einstein (and others) demanded that there be as few accidents in Nature as possible, that observational symmetries should manifest more fundamental, more deep-seated, symmetries.

It has to be admitted that the absolutist approach is often more intelligible than its relativistic counterpart. This is to my mind one important reason why we ought to be suspicious of absolutism. Speaking personally: despite my conviction that Relativity gives the correct account of physical processes, I still find it much more satisfying, while thinking about inertial motions, first to fix a universal time, then to lay down a unique medium, and finally to imagine infinite empty Euclidean boxes moving equably in the medium. It is also often more convenient, because more intelligible, to take length contractions and clock retardations as real physical effects caused by motions in the ether. It seems however that intelligibility is often a subjective quality arising from repeated use. The history of physics can be viewed as one continual struggle between intelligibility on the one hand and mathematical coherence on the other; with empirical progress being largely on the side of mathematical coherence. This is what Planck meant be saying that, as it evolves, science becomes both less anthropomorphic, less psychologically acceptable, and more objective.[3] Good examples of this process are: The Copernican Theory of earthly motion, Galileo's and Descartes' principles of inertia, Newton's actions-at-a-distance, relativistic space-time and Quantum Mechanics. In Quantum Mechanics we have a largely unintelligible, but mathematically coherent and empirically successful physical theory. This can be taken to indicate that Nature is objective – i.e. structured independently of our minds – precisely because it is so systematically unintelligible as thereby to become predictable. All we can say is that a certain mathematical formalism somehow, i.e. we do not exactly know how, faithfully reflects some aspects of physical reality. Going back to Mackie's position: to accept a philosophical argument which makes no difference to physics as a constituted system and moreover runs counter to successful physical practice is like taking opium: it makes one forget that only *empirically* successful programmes stand any chance at all of revealing something about the structure of *physical* reality. One would thus be accepting for metaphysics the role which the Vienna Circle allotted to it; a role similar to that of poetry or of certain forms of religion. Poincaré was I think the first philosopher to speak contemptuously of certain hypotheses which he termed *indifferent* because they act as props to our imagination without in the least affecting our scientific decisions.[4] I think one ought to be a little more ambitious on behalf both of metaphysics and of philosophy in general.

## NOTES

[1] Cf. H. A. Lorentz, *Theory of Electrons*, second edition, 1915, republished, New York, 1952, § §192–194 and Note 72.
[2] Cf. my article, 'Why did Einstein's programme supersede Lorentz's? (II)', *British Journal for the Philosophy of Science* **24** (1973), 223–237.
[3] *G. M. Planck:* 'Die Einheit des physikalischen Weltbildes' (1908), reprinted in H. Planck, *Vortäge und Erinnerungen*, Darmstadt, West Germany, 1973.
[4] H. Poincaré, *La Science et L'Hypothèse*, Paris, 1902, Chapter 9.

# TIME AND CAUSAL CONNECTIBILITY

LAWRENCE SKLAR

# PROSPECTS FOR A CAUSAL THEORY OF SPACE-TIME

What could possibly constitute a more essential, a more ineliminable, component of our conceptual framework than that ordering of phenomena which places them in space and time? The spatiality and temporality of things is, we feel, the very condition of their existing at all and having other, less primordial, features. A world devoid of color, smell or taste we could, perhaps, imagine. Similarly a world stripped of what we take to be essential theoretical properties also seems conceivable to us. We could imagine a world without electric charge, without the atomic constitution of matter, perhaps without matter at all. But a world not in time? A world not spatial? Except to some Platonists, I suppose, such a world seems devoid of real being altogether.

Given this sense of the primordialness and ineliminability of spatio-temporality, it is not surprising that many have tried, in one way or another, to account for or explain away other features of the world in terms of spatio-temporal features. Witness Descartes on extension as the sole real mode of matter and the Newtonian-Lockean program in general.

Yet there is a counter-current to this one. There is a collection of programs motivated by the underlying idea that either all spatio-temporal features of the world are to be reduced to other features not prima facie spatio-temporal, or at least that some of the spatio-temporal features of the world are to be eliminated from a fundamental account in terms of a subset of themselves and other features not prima facie spatio-temporal. Collectively these programs are sometimes said to espouse 'causal' theories of space-time. As we shall see there is a wide range of such programs with a variety of conclusions as their goals and with quite different underlying rationales. Some, indeed, probably ought not to be called causal theories at all.

What I would like to do here is to examine at least two classes of such programs, explore some examples within these classes, disentangle some of the confusions which have resulted from an insufficient awareness that such a diversity of goals and presuppositions exists in so-called causal theories of spacetime, and, finally, to suggest that for all the diversity both programs lead in the end to quite familiar and related kinds of philosophical perplexity.



*R. Swinburne (ed.), Space, Time and Causality, 45–62.*

## I

Both classes of causal theory I will discuss have as their members theories which allege that some or all spatio-temporal relations 'reduce to' or are 'eliminable in terms of' some class of relations, relations either not spatio-temporal at all or, at least, members of a modest subset of the original presumed totality of primitive spatio-temporal relations. In the theories of both classes it is alleged that the connection between associated 'reduced' relation and 'reducing' relation is more than a mere accidental coextensiveness. Indeed, more than lawlike coextensiveness will be claimed for the relations. In both cases the association will be declared necessary, although, as we shall see, the grounds for the allegation of necessity of the association (if not the kind of necessity) will be very different in the two cases.

But the motivation for alleging that the reduction ought to be carried out will differ markedly in the two cases. And the kinds of arguments which can be brought forward to either support the claim of reduction or attack it will differ as well.

What are the two classes of 'causal' theories of spacetime? Crudely the distinction is this: some 'causal' theories tell us that a spatio-temporal relation must be considered as reduced to a causal relation because our full and complete epistemic access to the spatio-temporal relation is by means of the causal relation, and therefore, some variant of a verificationist theory of meaning tells us that the very meaning of the predicate specifying the spatio-temporal relation is to be explicated in terms of the predicates expressing the causal relation. The other 'causal' theories tell us that we are to take the spatio-temporal relation to be reduced to the causal relation because it is a scientific discovery that the former relation is identical to the latter.

Let us look at some features of the epistemically motivated alleged reductions first.

## II

The idea that spatio-temporal relations must be identified with causal relations on epistemic grounds has its origins both in general empiricist and positivist claims about the association of meaning with mode of verification and, of course, in Einstein's critique of the notion of simultaneity for distant events. The basic line is that, just as Einstein showed us that we could only make concrete sense of the notion of simultaneity by identifying that relation with some specific causal relation among events, so in general a concept is legitimate if and only if a physically possible mode of verificaltion is associated with its application. This shows us that in general non-local spatio-temporal notions can only be understood by their identification with some appropriate causal relation among events.

There are a number of common features to causal theories of space-time of this sort which it would be well to point out.

First, there is a distinction drawn between those spatio-temporal features of the world to which epistemic access is taken as primitive. These relations remain in the reduction basis to which other spatio-temporal relations are reduced. Generally it is *local* spatio-temporal relations which have this unquestioned status. For example, in the Einstein critique of distant simultaneity, simultaneity for events at a point is never in question. It is simply assumed that such a relation exists and can be known to hold or not hold without the invocation of some causal means of verification. In this, and in other contexts as we shall see, the notion of what constitutes a continuous causal path in spacetime is also taken as primitive. In the Einstein 'definition' for distant simultaneity this appears in the implicit assumption that we can identify one and the self-same light signal throughout its history.

For spatio-temporal relations which don't fall into the primitive class, whatever that is taken to be, it is assumed that one must construct a meaning for such relations out of those available. Here it is frequently alleged, implicitly, that a distinction must be drawn between two sorts of defined spatio-temporal relations – those which have genuine fact-like status and those which merely attribute relations to the world reflective of conventions or arbitrary decisions on our part. For example, in the special relativistic case it is frequently alleged that local simultaneity and causal connectibility are facts about events in the world, but distant simultaneity is merely a matter of convention. Such arguments are usually backed up by claims to the effect that which events are simultaneous at a point and which causally connectible is not a matter of theoretical choice on our part, but that we could redescribe the world in such a way as to change the distant simultaneity relations among events leaving the genuinely observational predictions of our new theory unchanged from those of an older account which ascribed different simultaneity relations for distant events.

Built into these epistemically motivated reductions is an implicit distinction between a priori and empirically given epistemic limitations. In the Einstein critique, for example, the assumption that at least local simultaneity is directly given epistemically, and hence not up for epistemic critique, is one which is simply imported into the critical situation in an unquestioned way. What the ultimate roots of this assumption are remains in question. All that I am noting here is that within the context of this particular epistemically motivated critique, that this particular relation is 'given' to us simply is taken for granted a priori.

On the other hand, it is taken as an empirical matter, very much of interest in the context of the particular critique, as to just what the nature of the causal relationships are which allow us to extend outward from our primitive base of local spatio-temporal relations to a full set of spatio-temporal relations by means of causal definitions. For example, it is crucial to the Einstein critique that light be the maximally fast causal signal, that transported clocks not provide a unique synchronization, etc. Essentially, concepts are going to be admitted only if there can be associated with them a means of verification and, using the old positivist terminology, the limits of verification are going to be taken to be what is physically possibly verifiable, making the question of just what physical processes are lawlike allowed crucial for understanding our limits of legitimate concept introduction by definition.

Epistemically motivated causal theories of spacetime will generally attribute to some associations of spatio-temporal and causal relations a necessary status. Here the notion of necessity is familiar. If a spatio-temporal relation not in the primitive basis is to be introduced into our theory by means of a definitional association with some causal relation, then the meaning of the spatio-temporal relation is fixed by its definition and the connection of the spatio-temporal relation and the causal is one of analyticity and, hence, of necessity. We will have more to say of this shortly.

Finally it is interesting to note that in many such epistemically motivated causal theories of spacetime, the real set of primitive relations is plausibly taken to be itself a set of spatio-temporal relations. In the relativistic critique, for example, local simultaneity is plainly itself a spatio-temporal relation. But what about the all important notion of causal connectibility? When one realizes that the critique takes causal connectibility to be connectibility by a genidentical signal of the velocity of light or less, and that, at least plausibly, genidentity seems to be at least in part, and in crucial part, spatio-temporal continuity, one begins to see the motivation behind the claim that such 'causal' theories of spacetime might better be construed as epistemically motivated attempts to reduce the totality of spatio-temporal relations to a proper subset of themselves – in this case the subset of local simultaneity and continuity along a subset of one-dimensional spacetime paths.

## III

At this point it is useful to go a bit further into the dialectic which seems, inevitably I think, to ensue in the course of some such epistemically motivated reductions. The bulk of the debate generally hinges on such questions as the kind and nature of the connection between the accepted primitive elements and the defined concepts. Less attention has been paid, unfortunately I think, to the presupposition of the existence of a correct primitive set at all.

As an example of such a dialectic consider the recent debate on the 'conventionality' of the metric and of distant simultaneity in relativity. Following the usual presentations of special relativity it is frequently asserted that the metric of Minkowski spacetime and the relations of simultaneity for distant events are matters of convention, for they are not, in this usual presentation defined purely in terms of local simultaneity and causal connectibility.

But reference to the work of Robb, and more modern versions of the same results, shows us that we *can* offer definitions of spatial and temporal separation in terms of causal connectibility alone. Indeed, if we impose some weak constraints on the spatio-temporal relations, they are uniquely so definable. Does this then show that these relations are *not* conventional? [1]

Reference is then made to general relativity. Here it turns out that even in the cases where Robbian axiomatic constraints on causal connectibility are satisfied, that is, in spacetimes globally conformal to Minkowski spacetime, there will now be a divergence between spatio-temporal metrics as Robbianly defined and the usual metrics taken for the spacetime. And, in the general case, the Robbian axioms won't even be fulfilled, thus vitiating the very possibility of 'causal' definitions of the metric notions in the Robbian vein. [2]

These results show us a number of things. First, the coextensiveness of a 'causal' and a spatio-temporal relation will sometimes be a contingent matter. Which relations are coextensive with which others will depend upon just what kind of spacetime we live in. If we take it to be a conceptual and a physical possibility that the spacetime we are in is one in which the coextensiveness required for a particular causal theory of spacetime breaks down, then we will be able to attribute neither physical necessity nor 'analyticity' to the propositions asserting that the coextensiveness holds. Then, even if the spacetime in which we live turns out to be one in which the coextensiveness does hold, we will still view this as an inadequate relationship on which to build an epistemically motivated reduction of the spatio-temporal to the causal relation, for we will still view it as lacking the requisite necessity demanded of a reductive definition of this kind.

Exactly the same dialectic can be seen repeated in recent discussions of the possibility of a causal theory of spacetime topology. In Minkowski spacetime it is easy to characterize the manifold topology of the spacetime in terms of a causally defined topology – the Alexandroff topology. In spacetimes of general relativity the same definition will hold, provided that the spacetime is suitably non-pathological in the causal sense. In particular, the identification of the usual manifold topology with the Alexandroff topology will hold just in case the spacetime is strongly causal. If we weaken the causal constraint on spacetimes of strong causality, we can still show that in cases where the spacetime satisfies a somewhat weaker demand of non-pathology, past and future distinguishingness, the manifold topology will still be at least implicitly definable in terms of the notions of causal connectibility.

Where that weaker constraint is violated, however, it is possible to find topologically distinct spacetimes which are alike in their causal connectibility structure, indicating that one cannot, in these cases, even implicitly define the topology in any terms which take only causal connectibility as primitive. Once again we have the situation that a coextensiveness holds between a spatio-temporal relation and some causal relation in certain familiar spacetimes, but the conceptual possibility of other spacetimes, and perhaps their lawlike possibility as well, the usual laws of general relativity not precluding causally pathological spacetimes, shows that the correlation of spatio-temporal and causal relation is too weak to have the character of analytic necessity implicit in the epistemologically motivated reductionist program. [3]

These failures of the epistemically motivated program of providing causal definitions for spatio-temporal concepts ought not lead us to prematurely abandon the program. Consider, for example, the program to define the topology of spacetime in a causal manner. If we subtly change our notion of what is to count as the causal primitive in the defining base, a far more persuasive version of epistemically motivated causal definition can be constructed.

Instead of taking as one's causal primitive causal connectibility or variants of that notion, take instead the notion of a continuous causal path, in relativity the path of some suitable genidentical causal signal. One can show that in any general relativistic spacetime whatever, the set of continuous causal paths will certainly implicitly define the usual manifold topology. That is, if we have two such relativistic spacetimes, each endowed with a manifold topology, then any one-to-one mapping between them which preserves continuous segments of causal paths will preserve the manifold topology as well.

Even in this case we must be just a little cautious. Other, non-manifold topologies will still be compatible with the given set of continuous causal path segments, but not be homeomorphic to the usual manifold topology. It would be at this point that a defender of the causal theory we have in mind would likely move, with some plausibility, to the line that the difference between the usual manifold topologies and these new, non-standard, topologies, unlike that between two distinct manifold topologies, is merely a matter of conventional description of the same 'real' topological facts, these taken to be exhausted in our characterization of continuity along causal paths.[4]

IV

Reflection on the dialectic illustrated by the above examples shows us, I believe, a number of important things about epistemically motivated causal theories.

First, there is much we would need to say in justifying any such reduction concerning the status of the propositions which serve to define the spacetime notions in terms of the causal. Merely showing that in some particular spacetime there happens to be a coextensiveness between some spatio-temporal feature and some causal feature won't do to establish the definability of the former in terms of the latter. Not even lawlike coextensiveness will do, I believe. For the reduction to be acceptable to us we want the connection to have that kind of necessity traditionally associated with analyticity. We want to be convinced that, in some reasonable sense, the causal notion really captures 'what we meant all along' by the spatio-temporal notion.

Consider for example the critique of the Robbian causal definitions of metric notions in Minkowski spacetime. Even if we believed spacetime was Minkowskian, even if we believed that this was a matter of law, we would, I maintain be reluctant to accept Robb's 'definitions' as definitions of the spacetime metric notions. They would, we think, grossly misrepresent the verification procedures commonly associated with these notions and, for the familiar reasons associating meanings with primary conditions of warranted assertability, so misrepresent what we meant by the concepts. A definition for distant simultaneity which resorted, as Robb's do, to conditions of causal connectibility arbitrarily far out in the spacetime simply doesn't seem to capture what we meant by distant simultaneity, or rather the closest thing to what we meant by it prior to realizing the existence of the newly discovered constraints on the physical possibilities of verification, all along. But the Einstein definitions do have just that virtue lacking in Robb's. For that reason many would say that they correctly define the metric notions, even if this leads some who accept these definitions to submit to the claim that accepting them imposes the burden of relegating metric features to the realm of conventionality.

Accepting the above means, of course, accepting some sort of analytic-synthetic distinction among the propositions of one's total theory. Just how much of the positivist creed one need swallow here is another story, one I would rather not pursue here.

Next, it would be useful to reflect somewhat on the reduction basis presupposed by the epistemically motivated causal theories we have had in mind. In all of the cases at which we have looked, the same items in the reduction basis come up again and again: simultaneity at a point, causal connectibility and continuity along causal (timelight or lightlike) paths. Since the second can plainly be defined, in the relativistic context if not in general, in terms of the third, events being taken to be causally connectible just in case there is at least one continuous causal path containing them, we may focus on the first and third notion. While there may be some intrinsically causal aspect to the third, the first, local simultaneity, seems itself a spatio-temporal notion with no particular 'causal' nature at all. The third also on reflection looks more intrinsically spatio-temporal than causal. If being a continuous causal path means being the path of some genidentical particle, and if genidentity is unpacked itself in terms of spatio-temporal continuity, it seems that it is this notion of a continuous one-dimensional path in the spacetime which is primitive rather than anything specially to do with causality. Of course not all such paths are taken as primitive, only those of a timelike or lightlike nature, and these are, indeed, the paths, in the theory, of causal connection. But is it *that* aspect of them which leads us to place their continuity and discontinuity in the defining basis? Isn't it, rather, that they are the paths 'in principle' traversible by the experience of an observer, and, hence, the paths whose continuity or discontinuity could be determined in a 'direct', 'non-inferential' and 'theory independent' way? Perhaps the title of causal theory of spacetime is really a misnomer for the epistemically motivated theories we have been examining. On reflection they seem, rather, to be theories which attempt to reduce the total structure of spacetime to the structures characterizable in terms of a proper subset of our full set of spatio-temporal concepts, the reduction being rationalized by the claim that our full epistemic access being limited to such things as local simultaneity and continuity along causal paths, it is only these relations which we should take as bearing the full load of the 'real' nature of spacetime, the rest to be confined to the realm of, at best, the defined and, at worst, the merely conventional.

This brings us to our last, and most important, reflection on the structure of these epistemically motivated reductions. The fundamental problem with all such reductions is that they always seem to go too far. We would like to eschew absolute space with the neo-Newtonians, abandon global inertial frames with those who invoke curved spacetimes instead of flat spacetimes plus gravitational fields, and abandon a non-relativized notion of simultaneity with relativists. But the epistemic critique which allows us to do so, basically a unity with the kind of epistemically motivated causal theory we have been discussing, seems, when carried to its rational limit, to take us too far.

If we really are to confine ourselves in the reduction basis to the directly apprehended, non-inferential, non-theory loaded realm, and if that is not what characterized those privileged spacetime concepts what does single out their special status, we rapidly find ourselves slipping into the attribution of conventionality to many of our formerly most treasured spacetime features, to eschewing substantivalism for relationism, and for then continuing on to an account of the reality of spacetime which is at best phenomenalistic and at worst solipsistic.[5]

The problem is, of course, a familiar one, and one far more general than merely a problem with theories of spacetime. If we fail to make a distinction in kind between that which is epistemically accessible independently of hypothesized theory and independently of inference and that which is not, it becomes hard to see how we can carry out the epistemic critiques and attributions of theoretical equivalence we want to maintain. We *want* to say that Heisenberg and Schrödinger are merely two representations of one and the same quantum theory, we want to prefer, on Ockhamistic grounds, neo-Newtonian space-time to Newton's, curved spacetime to flat plus gravity and relativity to aether theories. Yet all the arguments in favor of these preferences rely, ultimately, on an observational/non-observational distinction of a more than merely contextual and pragmatic nature.

There are, of course, those who would deny the very intelligibility of the fundamental distinction here. The litany is familiar: all observation is theory laden, direct apprehension is a myth, facts are 'soft' all the way down, etc. But from that point of view it is impossible to see the point of a causal theory of spacetime of the kind we have been considering in the first place, and, indeed, hard to see what the point is of the epistemic critiques 'built in' to many of the most fundamental physical theories we do accept.

The problem is this: If we are, say, going to eschew an absolute notion of simultaneity for distant events, resting our disavowal on an Einsteinian critique, then we ought to pare our reduction basis down to those spatio-temporal and causal concepts which truly deserve our respect as being purged of conventionality. But then it is hard to see why we ought to stop at local simultaneity and continuity along causal paths if these notions are themselves construed as outside the realm of non-inferential direct access. And if it is relations in this realm which are the real reduction basis, then are they not relations in the realm of private experience, in the realm of phenomenal space, and not inter-subjectively determinable physical relations at all?

I have no intention whatever of resolving any of these perplexities here, nor even of surveying the familiar collection of solutions or of allegations to the effect that the perplexities are merely the result of typical philosopher's confusion. Here I only want to emphasize the inevitability of these questions once a causal theory of spacetime of the epistemically motivated sort is entertained. What we shall see shortly is that a related, if subtly different, problem arises when we follow out the other variety of causal theories of spacetime to their fatal limits.

## V

We saw that the term 'causal' used in characterizing the epistemically motivated causal theories of spacetime was, perhaps, a misnomer. While causal theories might play a role in characterizing the primitive observation basis, the fundamental nature of the relations in that basis was itself prima facie spatio-temporal, although, of course, the relations in the basis are a proper subset of the full array of spatio-temporal relations. The other approach to causal theories is also somewhat mis-designated as 'causal', but for different reasons.

The theories we are concerned with now are those which allege that some or all spatio-temporal features of the world can be considered reduced to some other features, themselves not prima facie spatio-temporal, in the same sense in which material objects reduce to their molecular and atomic constituents, light reduces to electromagnetic radiation, lightning reduces to atmospheric ionic discharge, etc. In other words, the claims will be that there is a reduction of the spatio-temporal to the not prima facie spatio-temporal which is a reduction by means of property identification.

One example of such a claim is, I think, the theory that the direction of time reduces to the (overall) direction of entropic increase. Since it is the direction of entropy increase, rather than the direction of causation, to which the direction of time is to be reduced, there is something misleading about calling this a causal theory. Indeed, in one such version of the theory, Reichenbach's, the direction of causation itself is 'reduced' to that of time and ultimately to that of entropic increase.[6]

The nature of this alleged reduction has been sometimes misunderstood, by myself among others. The fact that it is meant to be a reduction by means of scientifically established identification vitiates any objection to the proposed reduction which relies on some epistemic priority for the temporal direction as directly or more immediately perceived than the entropic. After all, there is a sense in which we know that there is light before we know that there are electromagnetic waves. Optics nonetheless reduces to electromagnetic theory and light waves are identical to a kind of electromagnetic wave.

The entropic theory of time direction is a curious sort of identificatory reduction. It is not some class of things which is being reduced to a class of things, but a relation which is being reduced to a relation. Indeed, most versions of the theory are even a little more peculiar, for it is only a part of a relation which is being reduced. In standard versions of the theory some temporal relations are simply presupposed as primitive. It is only the asymmetry of the relation which is ultimately to be accounted for entropically. Again the theory is subtle for other reasons. It is the *general* direction of time which is established entropically, and this only in a statistical way. Not every pair of events which bear the 'afterward' relation the one to the other need bear any entropic difference the one to the other. Indeed, entropy may be out of line entirely as characterizing either event. In Reichenbach's version of the theory entropy establishes time asymmetry for some pairs of events, and the asymmetry is projected onto other pairs by means which aren't entropic in nature.

The main point is this: Time direction is supposed to reduce to the 'causal' notion of direction of entropic increase. But the justification of the reduction is not like that which rationalizes the epistemically motivated reduction. No claim is made to the effect that our sole epistemic access into the asymmetry of time is through awareness, in a more direct sense, of entropic states and their relative magnitudes, nor is there some more direct awareness of the underlying statistical facts about randomness of microstates. Rather it is an empirical discovery that the relation in the world which 'underlies' our sense of the direction of time is that of the temporal direction of entropy increase.

Just as one would establish the truth of the claim that light waves are electromagnetic waves by showing that such an identification would account for all the observed and theorized features of light, so this theory is justified by an attempt to show that everything we take to be the case about the asymmetry of time (or, rather the asymmetry of the world in time), can be accounted for in entropic terms. Hence Reichenbach's assiduous attempt to show us that the fact that we have records and memories of the past but not of the future, that we take causation as going from past to future, etc. all have their origin in the prevalent increase of entropy of otherwise unmolested branch systems. Hence also his project, never carried out, of trying to show that entropic increase was also at the root of our subjective 'direct' awareness of the asymmetric relation in time of events of private experience.

I am not in the least, of course, maintaining that this theory can be successfully established. Only that if it would be it would be a reduction of a spatio-temporal notion to some other not at all like the earlier causal theories of the epistemically motivated sort. Without wanting to demand perfect symmetry for the analogy, this causal theory of time (or of time asymmetry) is to those discussed earlier much as the theory that tables are molecules is to the theory that tables are logical constructs out of sense-data.[7]

Another example of such an 'identificatory' reduction of spacetime relations to relations not prima facie spatio-temporal is even more dramatic in its claims. Unfortunately, it can't be described in any real detail, since, at the moment, it is rather a speculative hope than a genuine reductive program. The idea here is this: we start with a collection of events which are to be the results of measurements or quantum systems. We look for algebraic inter-relationships of the results of distinct measurements imposed by the quantum laws. We then hope to 'reconstruct' the spacetime structure on events by identifying it with some abstract structural inter-relations among measurements algebraically described. The topological, manifold and metric features of spacetime are dropped as primitives, only to reappear as 'identical' with the underlying combinatorial causal relationships.[8]

Notice that in both these examples and indeed, in any such attempted reduction by identification, there is no attempt to show that the spatio-temporal relations don't exist. In showing that light is electromagnetic radiation we aren't attempting to show that light does not exist either. It is just that the reduced entity or relation is identified with the reducing. What constitutes the *asymmetry* of the reduction is an interesting question, given the symmetry of the identity relation. In the case of the identification of things with arrays of atoms it is presumably the asymmetric part-whole relationship. In the case of the identification of light with electromagnetic radiation, it is presumably the greater generality of the reducing theory vis-à-vis the reduced. There are lots of electromagnetic waves which are not light, but all light is electromagnetic waves, for example. What would motivate the asymmetry in the case of the entropic theory of time asymmetry or in the case of the algebraic-combinatorial theory of spatio-temporal relations in general I won't pursue.

## VI

At this point we should note some general features of causal theories of spacetime which are meant to be identificatory reductions, and we should contrast those features with some we noted of reductions which have the epistemic motivation.

We saw how, in the cases of epistemically motivated causal theories, there was an interesting mixture of a prioristic and a posterioristic elements. What counted as a spatio-temporal or causal feature to be included in the reduction basis was determined, a priori, by our views regarding features of the world available to our cognition without inference or mediation of theory. But the extent and nature of additional meaningful spatio-temporal concepts, as well as, at least in part, the distinction between the factual and conventional portions of our spacetime picture, depended upon the possibilities available for verification procedures. Since these were usually taken to be possibilities in the mode of physical possibilities, it was then an a posteriori matter just how the reduction basis could be extended out by possibilities of verification into a full spacetime account. That 'afterness' for events in each other's lightcones is a 'hard fact' and that the simultaneity of distant events is 'merely conventionally defined', for example, indeed that any such notion of distant simultaneity must needs be inertial frame relative, these are conclusions which follow from the factual matter of the existence of a causal signal of limiting velocity.

There is less a priorism in the kinds of spacetime theories we are now exploring. Proponents of these theories are likely to argue that just as there is nothing in the least bit a priori about the fact that light waves are electromagnetic waves or table arrays of atoms, so it is a purely empirical matter that the direction of time is that of entropic increase or that the structure of spacetime in general is merely the algebraic combinatorial relations among actual and possible quantum measurements.

Nonetheless, it is still possible, even given this view, to demand of the propositions connecting spacetime structures to their reduction bases a *necessary* status. At least that is possible if one makes the familiar distinction between metaphysical necessity and epistemic a prioricity, and also goes along with recent popular claims that genuine identity statements are necessary in the metaphysical sense.

Consider, for example, the claim that the direction of time is that of entropic increase. Presumably we could not have found this out a priori. It certainly doesn't seem to be 'intuitively' true, nor does it seem analytic in the usual sense. Whatever we meant, all along, by the future time direction it doesn't seem too plausible to claim that we had in mind, all along, the direction of entropic increase – in the sense in which a phenomenalist would argue that in speaking of material objects all along we had in mind, really, merely talk about actual and possible sensory experiences.

Yet, could there be a possible world in which entropy decreased in time (pace all the usual qualifications and subtlties here)? Perhaps not. Just as it has been claimed that in any possible world water is $H_2O$ and light is electromagnetic radiation, perhaps a case could be made out that in any possible world the future direction of time just *is* the direction of entropic increase. And in any possible world, one might claim, the spacetime structure just *is* the same combinatorial structure on measurements it is in the actual world. Of course, the argument continues, just as there could be worlds where something other than water had all the features which we took in this world as the 'stereotype' or 'reference fixing features' of water in this world, so, presumably, there could be worlds where what presented itself to us in the way spacetime does in this world was something different. But there does seem something spurious about this. It is a matter I will come back to.[9]

## VII

Now one could argue against any particular identificationist causal theory of spacetime in many ways. But the argument which is most important for us is one which, if it is correct, allows us to dismiss in advance the possibility of the correctness of any such identificatory causal theory of spacetime. In the particular case of the direction of time it has been expressed with some forcefulness, if less clarity, by Eddington. We know, he says, what the nature of the afterward relation is like, directly and noninferentially, from our immediate experience. We also know what the relationship is like which is that of one state of a system having greater entropy than another. And we know that these two relations are 'utterly different'. Neither 'afterness' (Eddington speaks of 'becoming') nor having a more randomized microstate can be viewed as 'somethings I know not what' in a kind of Ramsey sentence way of looking at theories, a mode of being picked out sufficiently unspecific to allow us to find out that the state in question does indeed have some nature utterly unexpected. In Eddington's language, "We have direct insight into 'becoming' which sweeps aside all symbolic knowledge as on an inferior plane" [10].

The identificationist's likely reply to this move is also clear. We have a direct, non-symbolic knowledge of hotness and coldness also. If Eddington's arguments were correct, ought we not to say that hotness and coldness could not be states of higher or lower mean kinetic energy of micro-constituents? Yet that is exactly what those temperature states are. Of course, such a line would usually continue, there is 'felt hotness' and 'felt coldness', but these are mere 'subjective states of awareness of the observer'. Hotness and coldness in the object are more properly thought of as 'symbolically' known through their effects, and it is these objective features of nature which are identified with the micro-states in the usual reduction of thermodynamics to kinetic theory.

But now, in the light of all this, just how plausible do our identificatory reductions of spatio-temporal features appear? Presumably here too the refutation of the Eddingtonian will require that 'afterness' as immediately apprehended, or, in the more general case, all spatio-temporal features as we know them in our 'manifest' world of direct experience, are, as in the case above, to be 'sliced off' from the physical world and disposed of in that convenient receptable 'the subjective awareness of the observer'. The afterness of the world, or its spatio-temporality is then as distinct from felt afterness or sensed spatio-temporality as agitated molecules are from the *felt* sensations of hot and cold.

That we are pressed in this direction by identificatory causal theories of spacetime which attempt to say that spacetime is nothing but some non-prima facie spatio-temporal structure in the world has an interesting historical antecedent. Meyerson argued that the direction of science was always to eliminate diversity in terms of unity. One such program would be that which attempted to eliminate the sensed diversity of the world in terms of a reduction of all other features to the spatio-temporal. This would be the Cartesian program in any of its older or newer guises (of which geometrodynamics is perhaps the most well known current version). All there is are changing modes of extension. He pointed out that in such a reduction of all that is physical to modes of extension, it was inevitable that much of the directly sensed diversity of the world would be expelled from consideration in the reductionist program by being conveniently ejected out of the physical world and into the 'merely subjective' realm of private experience. Here we see that exactly the same pressures work when spacetime is to be identified as some structure not prima facie spatio-temporal, as hold when the not prima facie spatio-temporal is identified with modes of extension.[11]

But the pressures may be even more disturbing here. It is one thing to dump sensed color, felt heat and all the other 'secondary' qualities into the realm of the mental, reserving as objective only extension and its modes and all those other, 'symbolically known', features which can be identified with the modes of extension. It is, at least to some, even more disturbing to have to count spatio-temporality itself, at least as it is 'directly present' to us in our manifest world of awareness, as merely secondary, leaving in the actual world spatio-temporality only as a 'symbolically known' set of relations identical in nature with some relations among states of affairs not prima facie spatio-temporal at all.[12]

## VIII

What are the prospects of a causal theory of spacetime? As we have seen there are two quite different schemes one might have in mind when advocating a causal theory. One might be proposing a reduction of spatio-temporal relations to a proper subset of themselves (and perhaps some genuinely causal relations as well), motivating the reduction by an epistemic critique which, fundamentally, identifies meaning with mode of verification. Here there will be problems regarding the association of reduced with reducing motion, the familiar problems of justifying a claim to analyticity for some proposition. More crucially, the reductionist theory will seem to drive one to a reduction basis not in the objective world at all, leaving one puzzled as to how one's causal theory of spacetime could really be a theory of the physical world at all.

On the other hand, one might instead be claiming that, as an empirical matter of fact, the spatio-temporal features of the world are identical with other features not prima facie spatio-temporal. Here the fundamental difficulty will be that in carrying out the identification in a plausible way we will be required to shunt into the realm of 'mere subjectivity' the spatio-temporal features of the world as we encounter them in our immediate experience. Once again difficult questions regarding the relationship of what we immediately apprehend to what is objectively present loom.

There have been, of course, numerous attempts to relieve us of the two anxieties into which we have been led. Perhaps some are convinced of their success. For my own part I believe that I have yet to see an account which will: (1) do justice to the persuasiveness of the epistemic critique as it appears in functioning science, particularly in the critical evolution of spacetime theories in recent years; (2) do justice to the persuasiveness of identificatory reduction as a means of achieving unification of theory, including its persuasiveness in such contexts as Boltzmannian–Reichenbachian identifications of the future direction of time with the direction in time of entropic increase; and yet (3) which will neither trap us in a solipsistic prison nor remove from the world of objective reality such essential features as existing in space and time *as we know them*. The prospects for the success of a causal theory of spacetime are no better or worse than the prospects in general of an adequate account of the relation of the 'scientific' to the 'manifest' world.

## NOTES

[1] The Robbian results are contained in A. Robb, *A Theory of Time and Space* (Cambridge University Press, Cambridge, 1914), and *The Absolute Relations of Time and Space* by the same publisher, 1921. Related more modernly derived results are in E. Zeeman, 'Causality Implies the Lorentz Group', *Journal of Mathematical Physics* **5** (1964), 490–493. A clear presentation of the mathematical results, along with an attempted philosophical use of them, is in J. Winnie, 'The Causal Theory of Space-Time', in J. Earman, C. Glymour, and J. Stachel (eds.), *Minnesota Studies in the Philosophy of Science*, VIII, *Foundations of Space-Time Theories* (University of Minnesota Press, Minneapolis, 1977), pp. 134–205. See also D. Malament, 'Causal Theories of Time and the Conventionality of Simultaneity', *Noûs* **11** (1977), 293–300.

[2] See L. Sklar, 'Facts, Conventions and Assumptions in the Theory of Space Time', in the *Minnesota Studies*, volume cited above, pp. 206–274.

[3] S. Hawking and G. Ellis, *The Large Scale Structure of Space-Time* (Cambridge University Press, Cambridge, 1973), esp. Chapter 6. Also D. Malament, 'The Class of Continuous Timelike Curves Determines the Topology of Spacetime', *Journal of Mathematical Physics* **18** (1977), 1399–1404, and L. Sklar, 'What Might be Right About the Causal Theory of Time', *Synthese* **35** (1977), 155–171.

[4] E. Zeeman, 'The Topology of Minkowski Space', *Topology* **6** (1967), 161–170 describes the non-manifold topologies. On the determination of the metric by causal curves and the philosophical relevance of this see the items by Malament and Sklar cited immediately above.

[5] For some early, and not very satisfying, reflections on this see H. Reichenbach, *Axiomatization of the Theory of Relativity* (University of California Press, Berkeley, 1965), Introduction.

[6] The origins of the entropic theory can be found in L. Boltzmann, *Lectures on Gas Theory 1896–1898*, trans. by S. Brush (University of California Press, Berkeley, 1964),

pp. 446–447. The fullest account of the theory is H. Reichenbach, *The Direction of Time* (University of California Press, Berkeley, 1956). For some critical objections to the account see J. Earman, 'An Attempt to Add a Little Direction to "The Problem of the Direction of Time"', *Philosophy of Science* **41** (1974), 15–47. Some criticisms of my own, which I now think largely misdirected, are in *Space, Time, and Spacetime* (Univeristy of California Press, Berkeley, 1974), pp. 404–411.

[7] More on this is contained in L. Sklar, 'Up and down, Left and Right, Past and Future', *Noûs* **15** (1981), 111–129.

[8] Perhaps the fullest attempt at a reduction of spacetime to algebraically characterized relations among quantum measurements is D. Finkelstein's set of pieces on the 'spacetime code'. *Physical Review* **184** (1969), 1261; *Physical Review D* **5** (1972), 320; **5** (1972), 2922; **9** (1974), 2219, and D. Finkelstein, G. Frye, and L. Susskind, *Physical Review D* **9** (1974), 2231.

[9] On the necessity of identification statements see S. Kripke, 'Naming and Necessity', in D. Davidson and G. Harman (eds.), *Semantics of Natural Language* (Reidel, Dordrecht, 1972), pp. 253–255, and H. Putnam, 'The Meaning of "Meaning"', in K. Gunderson (ed.), *Minnesota Studies in the Philosophy of Science*, VII, *Language, Mind, and Knowledge* (University of Minnesota Press, Minneapolis, 1975), pp. 131–193.

[10] A. Eddington, *The Nature of the Physical World* (Cambridge University Press, Cambridge, 1928), Chapter V, 'Becoming'.

[11] E. Meyerson, *Identity and Reality*, trans. by K. Loewenberg (George Allen & Unwin, London, 1930), esp. Chapters XI and XII.

[12] The fullest attempt to overcome the difficulties incurred by scientific-identificatory reductions of time to causality of which I am aware is that of H. Mehlberg in his 'Essay on the Causal Theory of Time'. Originally published in 1935–37, in French, the work in now available in English translation in the posthumously published *Time, Causality, and the Quantum Theory* (Reidel, Dordrecht, 1980). Mehlberg accepts the dissociation of experienced from physical time which is, it would seem, an unavoidable consequence of such a scientific-identificatory causal account. He attempts to mitigate this consequence, unsuccessfully I believe, by invoking a 'universal' time which subsumes the subjective and the physical times. The frequent congruence of these times is alleged to be the result of psycho-physical parallelism of the realms of immediate experience and the physical realm of the brain. The doctrine, though, requires one to make sense of the notion of simultaneity of physical and mental event. Since this simultaneity can, it would seem, be neither a simultaneity in the sense of subjective nor in the sense of physical time, it isn't clear to me just how the doctrine can be coherently formulated. In the 'Essay' see esp. Part II, and most esp. Chapter X and the Supplement.

RICHARD SWINBURNE

# VERIFICATIONISM AND THEORIES OF SPACE-TIME

Professor Sklar's paper brings out very clearly the difficulties for theories which attempt to reduce spatio-temporal relations either to a subset of privileged such relations or to something apparently very different. I find myself in very general agreement with almost everything which he writes. But feeling that he raises problems rather than solves them, I would like to attempt something more ambitious. Both the problems which he raises, of the proper limits to verificationism and to property-identification, inevitably hang over all discussion of space and time and are raised by other papers at this conference. Space precludes my considering both problems, and so I shall confine myself to considering the general issue of verificationism. I shall consider how far verificationism is supported by plausible philosophical arguments, and then argue that the kind of verificationism supported by such arguments gives no support to a Robbian programme. I apologize for the fact that I shall take some time over very general philosophical discussion before I apply my results to space-time talk. My excuse is that scientific talk about space and time has been influenced by verificationist presuppositions for the past century, perhaps more than any other scientific talk; and it is important to clear up the extent of their philosophical justification.

## I

The general feeling which leads to verificationism is the feeling that we cannot have knowledge of that which is beyond the range of possible experience. But this very general empiricist feeling generates a whole galaxy of very different philosophical doctrines. One such doctrine is a doctrine which I shall call word-verificationism, but which is not a verificationist doctrine in the normal sense at all. This doctrine is the following. A sentence *S* has a truth-value if and only if it is a grammatically well-formed sentence, in which the referring expressions have a reference and the property-words (including relation-words) have a sense. A sentence is grammatically well-formed if it has the same structure as some verifiable sentence. Among well known kinds of grammatically well-formed sentences are subject-predicate sentences (such as 'the door is brown' or 'this chalk has a mass of 10 gms'),



*R. Swinburne (ed.), Space, Time and Causality*, 63–76.

relational sentences ('The lift has a velocity of 10 ft $s^{-1}$ relative to the Earth'), universal sentences ('all the chairs in this hall are ten years old') and conditional sentences ('if I drop it, it will break'). The referring expressions may be proper names or definite descriptions of individuals or indexicals (expressions such as 'this' or 'today' which purport to pick out individuals by their relation to the speaker). A referring expression has a reference if there is a procedure for determining what we are talking about, and that procedure does pick out an individual. 'Saturn' has a reference, because I can get you to look out of the window at night and point in a certain direction, and there is an object in the direction in which I point.[1] A property word has a sense if it occurs in verifiable sentences and makes a difference to their verification conditions. Thus 'red' has a sense because many sentences which contain 'red' are verifiable, and that word makes a difference to their verification conditions. 'This wall is red' is verifiable, and the word 'red' makes a difference because if we substitute a different word (e.g., 'brown') the new sentence would become verified under different conditions from the old sentence. So a sentence such as 'the nearest planet to the Earth of galaxy M33 has a mass twice that of the Earth', satisfies the word-verificationist's requirements for having a truth-value, because the sentence is a grammatically well-formed sentence; other sentences of that pattern are verifiable – e.g., 'the nearest marble to me has a mass twice that of the marble over there'; the names which occur in it have a reference – I can show you what I am talking about when I refer to 'M33' and 'the Earth'; and the property words occur in verifiable sentences. Word-verificationism allows the cited sentence to have a truth-value, even if in some sense the sentence is not itself verifiable.

Word-verificationism has in effect been espoused by those very many philosophers, including both Aquinas and Hume,[2] who have insisted that words must be empirically cashable if they are to be used in meaningful discourse. It is indeed an immensely plausible doctrine. A sentence of a certain form could not have a truth-value unless speakers could understand the kind of claim it was making (e.g., ascribing a property to an individual, or affirming a relation to hold between individuals or whatever), and how could they understand that unless they could be shown (or have explained to them in comprehensible terms) what it would be like for it to be true? And how could it be explained in other terms unless those other terms were ultimately cashable in experience?[3] Again a sentence referring to an individual could not have a truth-value unless speakers could understand which individual was being talked about, and how could they do that unless there was a procedure for determining the matter? And a sentence could not have a truth-value unless speakers could understand which properties (including relations) it is talking about, and how could they do that unless these can be exhibited in experience or defined in terms of ones which can? But once speakers do understand which individuals are being referred to and which properties are being ascribed how can they fail to understand what is being said by sentences?

## II

Word-verificationism makes whether a sentence has a truth-value depend, not on its own verifiability, but on the verifiability of similar sentences. But word-verificationism is not verificationism proper, and that is a much more contestable doctrine. Verificationism proper claims that a sentence has a truth-value if and only if it can itself be verified. There are many varieties of verificationism proper, according to how we understand the 'can' – as denoting logical, physical, or practical possibility – , and the 'verify' – whether the verification is supposed to be incorrigible, to be 'beyond reasonable doubt', or to be simply confirmation, and by whom (at least one person, or everyone), and when (now, or at some time or other) the verification is supposed to be performable. The stronger your verificationism, the more sentences it rules out, which seem at first sight to have a truth value. If you insist on incorrigible verifiability by everyone, sentences about the mental life (such as 'I am in pain') fail to have a truth value. If you insist only on the logical possibility of verification beyond reasonable doubt, still all universal lawlike claims such as 'all light travels *in vacuo* at 300 000 $km\,s^{-1}$ relative to all inertial frames' fail to have a truth value. They make claims about going-on in spatio-temporal regions far distant from our own, and no agent could ever establish beyond reasonable doubt that they hold in those regions. To give such sentences a truth-value you need to insist that, whether the possibility of verification be practical, physical, or logical, the verification which must be possible is only confirmation. We can have evidence for such a universal sentence, although not evidence which puts it beyond reasonable doubt.

Two movements of recent years have given prominence to verificationism. The first and most evident one was of course the logical positivism of Carnap and Ayer. But, after a period of temporary decline, verificationism has come to the fore again in the 'anti-realism' of Michael Dummett and his followers.[4] As the latter have produced more sophisticated arguments for verificationism than the former did, let us look at those arguements to see what kind of support they give to verificationism. Dummett acknowledges the obvious fact which is central to the word-verficationist's arguments, that we learn from experience of verified sentences to construct new sentences which we do not immediately verify, but he claims that:

The process by which we come to grasp the sense of statements of a disputed class, and the use which is subsequently made of these statements, are such that we could not derive from it any notion of what it would be for such a statement to be true independently of the sort of thing we have learnt to recognize as establishing the truth of such statements.[5]

This sentence contains in embryo Dummett's two central arguments. The first of these arguments, the acquisition argument, is that since we derive our knowledge of what it is like for sentences to be true or false from situations in we have experience of the circumstances which make it justifiable to assert or deny them, our understanding of their truth and falsity must be an understanding which connects truth to justified assertibility, and falsity to justified deniability. This argument does not however seem very satisfactory. Certainly we gain our understanding of certain sentences from experiencing the circumstances in which they are justifiably asserted or denied, but it does not follow that our understanding of the new sentences which we subsequently learn to construct is such as to tie their truth or falsity inseparably to their justified assertibility or deniability. Perhaps our language can spread its wings beyong the circumstances essential for its birth.

Dummett, I think, lays much more stress on the second argument, and certainly much more force attaches to it. This second argument, the manifestation argument, claims that to know the meaning of a sentence is a practical ability, an ability to do something with it, an ability to manifest that knowledge in certain ways. For some sentences, knowledge of the meaning of the sentence is explicit knowledge – it is to be able to state what it means. But a speaker will have to state what the sentence means in terms of other sentences, the meaning of which he knows in some other way. (Otherwise we should have a set of circular definitions which never latched on to the world.) And what is it for the speaker to know what these other more primitive sentences mean? This is implicit knowledge manifested in recognising when it is correct to assert or deny a sentence. If a speaker can do that (in virtue of his recognition of what are the parts of the sentence, and how it is put together), he knows what a sentence means. And if he cannot do that, what on earth can his knowledge of meaning consist in? Knowledge of meaning is therefore basically a recognitional capacity. So we cannot understand what it would be like for sentences to be true or false except in so far as we understand when they are justifiably assertible or deniable.

Now clearly there are sentences with a truth-value which are never in fact justifiably asserted or denied by speakers. 'There is a large rock under this tree' may be true even though no one has ever checked it out or will ever check it out. That does not worry the anti-realist. The sentence has a truth-value because if circumstances arose when a hole was made under the tree and speakers looked in the hole, they would have justifiably asserted or denied the sentence. All that is required for actual knowledge of the meaning of a sentence is an ability if relevant circumstances were to arise, justifiably to assert or deny the sentence in those circumstances. The evidence that a man has this ability will be provided by his use of the words of the sentence in question on other occasions and by his use of verifiable sentences of the same pattern. That is, the satisfaction of the demands of word-verificationism (with respect to sentences which are verifiable in this way) is evidence that if appropriate circumstances were to arise speakers would respond appropriately.

But in that case are not all the sentences about which anti-realists worry ones which if speakers were put in appropriate circumstances they would correctly assert or deny? All sentences about the distant past or distant future are ones which if the speakers were put in appropriate circumstances, they would correctly assert or deny. The trouble is that they cannot be put in those circumstances. However if the anti-realist insists that the circumstances must be in some sense ones which can occur, the question arises – in what sense 'can' occur? In a deterministic universe it is not physically possible that anything could happen other than does happen. Now maybe our universe is such that if any man seeks to look under some large tree, something prevents him from doing so. Some farmer is predetermined to stop him or some bull to chase him away. In that case, 'there is a large rock under this tree' not merely will not be, but cannot be justifiably asserted or denied. Perhaps our universe is not a deterministic one, but we can hardly hold that whether our sentence has a truth-value depends on whether or not it is. So since 'there is a large rock under this tree' quite obviously has a truth-value even if speakers in the sense described are unable justifiably to assert or deny it, we may conclude that the mere physical impossibility of the conditions being realized in which speakers would justifiably assert or deny a sentence does not deprive the sentence of a truth-value, nor (in view of the anti-realist's general arguments) does it prevent speakers from now having an ability justifiably to assert or deny it in appropriate circumstances (if, per impossible, these were to arise). Circumstances may make it impossible for abilities to be manifested, but the abilities may exist all the same. I may have the ability to run a mile, or climb a tree, but be prevented from manifesting it by being locked up in prison. The point is that to have an ability is to be so constructed that in certain circumstances (if they were realized), you would (or would if you chose) do certain things. And you can be so constructed whether or not the circumstances can ever be realized.

It may be claimed that the ability which must exist if a sentence is to have a truth-value is not just the ability to assert or deny the sentence in appropriate circumstances, but the ability to bring about such circumstances. Implicit knowledge of the meaning of a sentence, Dummett sometimes claims, is a matter of the speaker's having an "effective procedure which will, in a finite time, put him into a position in which he can recognize whether or not the condition for the truth of the sentence is satisfied"?[6] But then what is his having an 'effective procedure'? Is it that if he chooses, he will get himself in a position to check that the sentence is true? But that seems false. There are many sentences for which we have no such effective procedure, which only luck will get us into a position to check and yet which intuitively seem quite clearly to have a truth-value. For example, all claims for which we do not currently have the scientific expertise to check out. Recall the famous sentence of *Language, Truth and Logic*, first published in 1936, 'there are mountains on the other side of the moon'.[7] Ayer gave this as an example of a sentence which was 'in principle' verifiable, although not in practice. No one then had the technique of travelling to the moon so as to view its other side, and for all they knew it might not have been physically possible so to do. And yet the sentence would have been 'in principle' verifiable, in the sense that if speakers were put in appropriate circumstances they could have verified it. And in general for all unrestricted existential generalizations of the form 'there are $\phi$'s somewhere in the Universe', we may have no effective procedure for determining truth-value; luck alone may enable us to do this, and yet that does not deprive those sentences of truth-value.

An anti-realist may claim that, even if it is not physically possible, it must be logically possible for the circumstances to be realized in which speakers can check out the truth-value of a sentence, if that sentence is to have a truth-value. But even that claim is not justified by his arguments. For granted knowledge is a recognitional ability, abilities may exist even if it is logically impossible for the circumstances to occur in which they can be manifested. I have the ability to recognize when someone has trisected an angle with ruler and compass, or found an even number which is not the sum of two primes – even though (I reasonably assume) neither task is logically possible. I have the ability because if a man did these things I would recognise that he had done so. I conclude that all that the anti-realist arguments show is that for a sentence to have a truth-value, it is necessary that speakers be able justifiably to assert or deny it if they were put for appropriate circumstances (i.e., speakers will so assert or deny, if they choose to seek to express the truth).

But this will inevitably be so with respect to any sentence for which the demands of word-verificationism are met. For in this case the sentence will make a claim of a kind which speakers know how to verify – e.g., (being a subject-predicate sentence) it will attribute a property to an individual, or (being a universal sentence) it will claim that all individuals which have one property have a certain further property. So speakers will know with respect to the given sentence the kinds of procedures appropriate to verifying it. These procedures will depend, more specifically, on which individuals are referred to and which properties, are designated in the sentence. But, ex hypothesi, the referring expressions do have a reference and the property words do have a sense and speakers know how these make a difference to the verification conditions of sentences. Hence speakers will know with respect to a sentence which satisfies the demands of word-verificationism, what (if it were to occur) would verify that sentence. So this weak form of verificationism proper, which is all that is justified by the anti-realist's argument, establishes the same sentences as having a truth-value as does word-verificationism.

These arguments yield a form of verificationism which does not insist even on the logical possibility of verification, only on the actual ability of speakers to verify in appropriate circumstances (the occurrence of which may not even be logically possible). For a sentence to have a meaning at a time speakers must have the ability at that time.[8] If the sentence is a sentence of a public language, more than one speaker must have this ability. Not believing in the possibility of any sentence being verified incorrigibly – for well-known reasons, I suggest that we must choose between insisting on verification 'beyond reasonable doubt' and mere confirmation.[9] But then there is quite clearly a continuous quantitative graduation between these two. We can have some evidence for a sentence or a bit more, or a bit more still, and if we have enough that puts the sentence beyond reasonable doubt.

Any attempt to insist on verification 'beyond reasonable doubt' would lead to a most arbitrary division. I suggest that any degree of verification, viz. mere confirmation, suffices. In this way as in other ways, we avoid Sklar's slippery slope (his p. 53) – here, as in most other places, the way to deal with slippery slopes, is not to get started on them. However, which ever kind of verification we select for word-verificationism, we shall naturally seek also for our weak form of verificationism proper.

III

What follows from all this for 'causal' theories of space-time? These theories seek to reduce talk of distant simultaneity to talk of local simultaneity and causal connectibility (or perhaps causal continuity), on the grounds that we have epistemological access to the latter alone. I do not think that the causal theorist need hold, as Sklar suggests (his p. 51) that "in some reasonable sense, the causal notion captures 'what we meant all along' by the spatio-temporal notion". The causal theorist may hold that quite a lot of what 'we meant all along' lacked truth-value (because of the impossibility of verification of his preferred kind); but maintain that the 'causal notion' captures the significant parts of our previous claims.

In order to hold that, he must hold that claims about absolute simultaneity at a distance have no truth-value. He must claim, with respect to the traditional example, from Special Relativity, that if a light signal is sent from Earth to Jupiter at 10.00 a.m. by an Earth clock and immediately reflected from Jupiter so as to arrive back on Earth at 10.10 a.m. by the Earth clock, there is no reading on the Earth-clock with which the event of the reflection from Jupiter is simultaneous absolutely. One can lay down conventional standards for simultaneity at a distance, i.e., a frame of reference $F$ and a light signal method for synchronizing clocks within it, and then say that relative to $F$ the reflection was simultaneous with the Earth-clock reading 10.05 a.m. (The use of the light signal method in $F$ involves assuming that light has the same one-way velocity in all directions in $F$, and so that the event of its reflection is midway in time between the event of its emission from and the event of its return along the same path to a given point. This assumption enables clocks to be synchronized at points throughout $F$.) But one could equally well choose another frame $F'$ (moving relative to $F$), and relative to $F'$ the reflection would be simultaneous with the Earth-clock reading 10.07 a.m. There is no absolute simultaneity, the traditional argument put forward by Einstein claims [10], because it is physically impossible to verify any claims about which reading on the Earth-clock between 10.00 and 10.10 is simultaneous with the reflection.

The physical impossibility of verifying 'beyond reasonable doubt' which event is simultaneous with the reflection arises from the fact that light is the fastest signal. One cannot send a faster signal from Earth to Jupiter to be reflected simultaneously with the light signal, which would leave Earth after and arrive back at Earth before the light signal and which would thus narrow the period on the Earth clock during which the reflection must have taken place.

Although one cannot conclusively verify such a simultaneity claim, one might suppose that one could obtain for it some more indirect confirmation. Simplicity is evidence of truth. Among theories which predict observations, we believe that one to be most likely to be true which is the simplest theory which predicts our observations. Perhaps one claim about absolute simultaneity is compatible with a simpler theory which accounts for our observations, than is some other claim. Not so, however, so long as we synchronize our clocks by the use of the light-signal method in some inertial frame, our rival claims about simultaneity are equally compatible with the simplest theory of our observational data – the Special Theory of Relativity. There is no detectable preferred frame of reference.

However, there is nothing inconsistent with any experimental data or with the basic theoretical structure of Special Relativity in supposing that there is an (undetectable) preferred frame, e.g., that the use of the light signal method in $F$ alone yields the true judgments of absolute simultaneity. Experiment shows only that light has the same two-way velocity in all frames of reference. If in $F$ it has the same one-way velocity ($c$) in all directions, then in $F'$ it will travel with a velocity less than $c$ in one direction, and a velocity greater than $c$ in the reverse direction. Nor would such a supposition upset the basic theoretical structure of Special Relativity.[11] The Lorentz Transformations remain. The transformation

$$t' = \frac{1}{\sqrt{1 - v^2/c^2}} \left(t - \frac{vx}{c^2}\right)$$

is to be read as stating the relation between the time interval $t'$ which would be measured in $F'$ if one assumed that in $F'$ light has the same one-way velocity in all directions, and the interval $t$ which would be measured if one made that assumption about $F$. Likewise the Minkowski geometry remains. Events are still time-like connectible or space-like connectible.

But 'non-time-like connectibility' is to be read as 'not linkable by a causal chain', not as 'not in any absolute respect temporally before or after'. The equations and the geometry remain; it all depends how you interpret them.

But as we saw, because of the physical impossibility of determining which distant events were absolutely simultaneous, Einstein and others wished to deny that there was any absolute simultaneity. Yet if my earlier arguments are correct, the mere physical impossibility of verification is not good grounds for asserting lack of truth-value. For speakers have the ability with the respect to the sentence 'the reflection was simultaneous with the reading 10.05 on the Earth clock' to recognize that it is true or false under the circumstances (which are no doubt physically impossible ones) that a signal of infinite veolocity is sent from Jupiter to Earth at the time of the reflection. Maybe there are more claims about the world that men can understand than claims whose truth-value they can discover.

An objector might agree that the mere fact that it is not physically possible for the circumstances to occur in which a particular sentence will be verified or falsified does not render it empty of truth-value. But he may claim that there is a crucial difference between the down-to-earth example of the rock under the tree, through which in part I made this claim plausible, and the case of absolute simultaneity. For, he may urge, the physical impossibility of verifying whether there is a rock under the tree arises from both the initial conditions (initial positions of observers, farmers, bulls, etc.) and from the laws of nature (how farmers and bulls affect observers). Whereas, he may say, the laws of special relativity alone (embodying the claim that $c$ is the fastest possible two-way velocity of a signal) prevent the verification of the claim of absolute simultaneity of two distant events $E_1$ and $E_2$. But the distinction is a spurious one. For in the case of any claim that two events $E_1$ and $E_2$ are simultaneous, when we are given no further information about $E_1$ and $E_2$ it is not laws of nature alone which prevent verification of claims of absolute simultaneity. It is also the fact that the events occur at certain distances from each other at certain times on their clocks synchronized by certain procedures. It is the combination of laws and set-up which prevents checks of absolute simultaneity. Of course if we include this description of the set-up into the description of the events, and describe them as events of such and such a kind, then laws of nature alone determine that the circumstances cannot be realized in which checks of absolute simultaneity between those events can be made. But it is just the same with the rock. If we describe the event of the uncovering of ground under this tree as the uncovering of ground under a tree of such a kind in a universe which began with an initial distribution of matter of a certain kind (no doubt, one picked out by a very complicated description) then laws of nature alone determine that such events cannot occur; and so that whether there is a rock under the tree is unverifiable. I conclude that the tree-under-rock example is no different in kind from that of distant simultaneity. If it leads us to accept that a sentence may have a truth-value even if it is not physically possible that it be verified, it ought to lead us to the same view with respect to a sentence affirming the absolute simultaneity of distant events.

Anyway, as regards our Universe rather than the matter-free Universe of Special Relativity, it seems to me practially possible in the weak sense of 'confirm', to verify such simultaneity claims. For although in the world of Special Relativity, there are no good grounds for preferring one frame of reference as the frame in which to make measurements rather than another moving with uniform velocity relative to the former, there are such grounds in our Universe. The laws of overall behaviour of matter in the Universe (viz. the laws of cosmology) take their simplest form when measurements are made relative to the 'fundamental particles' of cosmology, i.e., those frames of reference in the vicinity of a galactic cluster which have the mean motion of matter in the vicinity of that cluster. Simplicity is evidence of truth. The resulting cosmological theory yields a unique 'cosmic time' throughout the Universe. Adopting this has the consequence that local measurements made relative to the local fundamental particle rather than relative to any other local frame would be the true measurements. Hence we should assume that light has the same velocity in each direction relative to the frame stationary relative to the mean motion of matter in the vicinity of our galactic cluster. This will yield a unique value for the reading of the Earth-clock simultaneous with the event of reflection.[12]

Of course all this provides only weak confirmation of claims about distant simultaneity. And some anti-realists might attempt to tie possession of truth-value to the existence of stronger forms of verification (although I cannot see anything in their arguments which justifies such a tie). But a 'causal theorist' cannot afford to insist on stronger forms of verificationism. For the causal theorist analyses spatio-temporal claims in terms, not of 'causal connection' but of 'causal connectibility', i.e., the physical possibility of causal connection. And how does the theoriest know which events on Earth are causally connectible with the event of the reflection on Jupiter? By a complicated process of extrapolation from observation – of the velocity of light relative to different frames (as in the Michelson–Morley experiment), of the increase of mass with velocity – to the simplest theory which will account for these, Special Relativity. Unless the indirect process of extrapolation from data to remote but simple theories is legitimate, the causal theorist cannot have his allegedly epistemologically primitive data for the construction of his theory. But if the process is legitimate, it can be carried a bit further to embrace the data of cosmology as well, and to build from those data an account of preferred frames and confirmed assumptions about velocity.

My conclusion is that verificationist arguments provide no good grounds for adopting a causal theory of time.[13] Observations of causal connection (and of local simultaneity) indeed provide our grounds for asserting temporal judgements. But there may be temporal truths (e.g., about the simultaneity of distant events) which it is physically impossible for us to discover. Superficially there is more to simultaneity than causal connectibility, and verificationist arguments are inadequate to show otherwise.[14] A satisfactory science ought to reveal the extent of our ignorance, not pretend that what is not knowable is not true. Further, observations about causal connection (and local simultaneity) may provide confirmation for judgments of distant simultaneity; and the 'causal theorist' is in no position to insist on a stronger kind of verification.

## NOTES

[1] I do not investigate more generally the issue of what are the proper procedures for determining what we are talking about – e.g., whether proper names are to be elucidated in terms of a descriptive or causal theory.

[2] Thus Hume held that words which purport to denote 'ideas', i.e., in modern terms properties, do so only if we have had 'impressions' of them, i.e., observed instances of them. (D. Hume, *An Enquiry Concerning Human Understanding*, ed. by L. A. Selby-Bigge, second edition, Oxford, 1902, p. 22). Aquinas quotes Aristotle with approval for holding that "our knowledge begins from the senses" (*Summa Theologiae*, Ia. 84.6), and he held that the meanings of predicates applied to God "are known to us solely to the extent that they are said of creatures". It has been the recent programme of Donald Davidson to show how the meanings of compound sentences derive from their constituents. See, for example, his 'Truth and Meaning' in J. W. Davis *et al*. (eds.), *Philosophical Logic*, Reidel, Dordrecht, 1969.

[3] The demand that the sentence be grammatically well-formed may be understood in a wider or a narrower way. In the wider way, the demand is simply a demand that the sentence have the same syntactical structure as some verifiable sentence. Thus if a given verb only makes a verifiable sentence when followed by an object-term, then it must be followed by an object-term in any sentence which is to be grammatically well-formed. Thus 'John opens' is not grammatically well formed, because 'opens' is always followed by an object-term in verifiable sentences. We need to say 'John opens the door'

or 'John opens a book', to have a sentence with truth-value. This kind of grammatical formation is the kind with which grammarians deal. Yet it allows such sentences as 'Saturday opens the door' or 'truth has four equal sides' to have truth-value. Many philosophers have wished to deny truth-value to such sentences, claiming that they are 'meaningless' because they commit some sort of category mistake. To form a sentence with a truth-value it is no good inserting any referring expression into the gap in '______ opens the door'. You have to insert a referring expression which picks out a substance, maybe even necessarily a living substance, not a period of time. So 'John opens the door' or 'the dog opens the door' have truth-value, whereas 'Saturday opens the door' does not. To insist that similarity of form to that of verifiable sentences includes similarity in respect not merely of syntax but of categories, yields a more restrictive form of word-verificationism. To spell it out, we would need a satisfactory philosophical doctrine of categories – i.e., a doctrine of the kinds of thing which can exist, and the kinds of property which things of different kinds can possess. ('Can' here is the logical 'can'. Our concern would be with the logical possibility of different kinds of thing possessing properties of different kinds.) Yet the restriction seems to me an unnecessary one. It seems not unnatural to say that 'Saturday opens the door' is not meaningless, like a collection of mumbo-jumbo expressions or even like 'John opens', but comprehensible and false. However, in either form word-verificationism is a doctrine which makes whether a sentence has a truth-value depend, not on its own verifiability, but on the verifiability of similar sentences.

Oddly, unknown to me when I was writing my paper, my cosymposiast had just published a paper on the doctrine which I call word-verficationism, basically sympathetic to it but exhibiting the difficulties of spelling it out, such as those described above. See Lawrence Sklar 'Semantic Analogy', *Philosophical Studies* **38** (1980), 217–34.

[4] See Dummett's papers 'Truth', *Proceedings of the Aristotelian Society* **59** (1958–9), 141–62; 'The Reality of the Past', *Proceedings of the Aristotelian Society* **69** (1968–9), 239–58, and 'What is a Theory of Meaning (II)', in G. Evans and J. McDowell (eds.), *Truth and Meaning*, Oxford 1976. All are included in the collection of his papers *Truth and Other Enigmas*, London, 1978. Sometimes Dummett seems to argue impartially between the realist and the anti-realist, sometimes to come down explicitly in favour of the anti-realist. A more explicit recent anti-realist is Crispin Wright. See his 'Truth Conditions and Criteria', *Proceedings of the Aristotelian Society*, Supp. Volume, **50** (1976), 217–245: 'Realism, Truth-Value Links, Other Minds, and the Past', *Ratio* **22** (1980), 112–132; and the passages about anti-realism in his recent book *Wittgenstein on the Foundations of Mathematics*, London, 1980 (especially Chapters 10, 11 and 12).

[5] 'The Reality of the Past', pp. 243f.

[6] 'What is a Theory of Meaning (II)', p. 81.

[7] A. J. Ayer, *Language, Truth and Logic*, second edition, London, 1946, p. 36.

[8] Even if we insist on the possibility of verification 'beyond reasonable doubt', it will not follow that sentences about past or future lack truth-value. For a sentence about the future (or past) will normally be such that a speaker knows how to verify them now if the appropriate circumstances are now present (which may not be in some sense logically possible).

[9] Dummett ties verificationism to the possibility of 'conclusively' establishing or falsifying a sentence. See 'What is a Theory of Meaning (II)', p. 114. It seems natural to understand 'conclusively' as 'beyond reasonable doubt'.

[10] "So we see that we cannot attach any *absolute* signification to the concept of simultaneity, but that two events which, viewed from a system of coordinates are simultaneous, can no longer be looked upon as simultaneous events when envisaged from a system which is in motion relatively to that system". A. Einstein 'On the Electrodynamics of Moving Bodies', in H. A. Lorenz *et al.* (eds.), *The Principle of Relativity*, London, 1923, pp. 42f.

[11] For the beginnings of a detailed working-out of the consequences of supposing there to be a preferred frame of reference, and demonstration of its consistency with the predictions of Special Relativity, see A. Grunbaum, *Philosophical Problems of Space and Time*, second edition, Dordrecht, 1973, Chapter 12, Section B. For some development of these issues, see also his Chapter 20; Philip L. Quinn, 'The Transitivity of Non-standard Synchronisms', *British Journal for the Philosophy of Science* **25** (1974), 78–82; and R. Francis, 'On the Interpretation and Transitivity of non-standard Synchronisms', *British Journal for the Philosophy of Science* **31** (1980), 165–173 (noting especially footnote 4).

[12] I summarize here an argument given more fully in my *Space and Time*, second edition, London, 1981, Chapter 11.

[13] It does not seem evident to me that verificationism is the main motive behind Winnie's attempt at a causal theory, although Sklar cites him in a verificationist context. See Winnie's remarks on p. 190 (and also p. 187) of John A. Winnie, 'The Causal Theory of Space-Time', in J. Earman *et al.* (eds.), *Foundations of Space-Time Theories*, Minnesota Studies in the Philosophy of Science, Vol. 8, Minneapolis, 1977.

[14] Nothing in this paper should be taken as ruling out the possibility that different arguments from those of the verificationist might show that the concept of absolute simultaneity had no application to distant events. The second kind of reductionist whom Sklar considers might have such arguments. There is also the interesting argument given by Dr. Zahar in his contribution to this volume, that supposing there to be absolute simultaneity involves supposing a very subtle correlation between velocities as measured in different frames (by the light-signal method in each frame), of just the exact value needed to prevent us from ever discovering which events are absolutely simultaneous – a sort of conspiracy on the part of nature. See p. 39 of his contribution. However that argument is an argument to show that sentences affirming distant events to be absolutely simultaneous are false, not that they lack truth-value (which is the verificationist's claim).

# TEMPORAL AND CAUSAL ASYMMETRY

RICHARD A. HEALEY

# TEMPORAL AND CAUSAL ASYMMETRY

## I

There is a sense in which the causal relation is associated with the temporal direction earlier to later rather than the reverse. There is also a sense in which irreversible natural processes define a unique temporal direction. Just what these senses are needs to be clarified. But pending such clarification it seems clear that one can raise the question whether there is any relation between these two directions (the "direction of causation" and the "arrow of time"), and if so, what it is. This paper attempts to clarify, sharpen and return a provisional answer to this question. The philosophical aim which lies behind the investigation (but sometimes becomes more explicit) is to approach an improved understanding of the place of causation in physics, in the world, and in our thought. To introduce the relevance of our question to this aim, consider two opposing approaches to causation.

The physicalist approach to causation proceeds from a challenge. Either there is something in the physical world which corresponds to causation, or our talk of causation is empty and should be proscribed. To determine what, if anything, there is to causation, we should therefore study descriptions of the contents of the world generated by our best physical theories. If we find anything there corresponding to our prior understanding of causation then we have legitimized causal talk by finding out what causation really is: if we do not, then we have demonstrated the illegitimacy of our causal talk.[1]

The conceptualist approach to causation stresses the historical development of our causal concepts and their central role in our thought. A hypothesis in philosophical anthropology is advanced, according to which our causal concepts trace their origins to the primitive human experiences of pushing, pulling, and more generally producing changes in objects in the environment (including other humans). Our contemporary causal concepts are held to have developed from such primitive origins, becoming extended and modified in their scientific employment to apply to processes in the inanimate world, but still retaining the connection with human agency that causes are potential means by which humans could, at least "in principle", bring about their effects.

Now suppose we describe a certain counterfactual situation and then



*R. Swinburne (ed.), Space, Time and Causality*, 79–103.

pose a question about it to both physicalists and conceptualists. The situation is as follows: the entire contents of some large four-dimensional region of spacetime are to be imagined to be "removed from" the spacetime, "inverted in" the dimension, and "reinserted into" the spacetime, with the result that all processes in that region of spacetime now appear to "run backwards". (Either we suppose that the region of spacetime can be chosen so that no physical laws are consequently violated at the boundaries or we simply ignore any such spatiotemporally distant violations. We further suppose that the physical laws within the portion are time-reversal invariant so that none of them are "subsequently" violated within the region.)[2] We then ask 'Do causes now succeed their effects within this region where "prior to the inversion" they preceded them?'

The non-sceptical physicalist may be supposed to have identified some physical features in the region "prior to inversion" as corresponding to causation. If he follows the same principles of correspondence he will apparently conclude that after the imagined "inversion" the temporal direction of causation in this region of spacetime is reversed: and so the question receives the answer 'yes'. The conceptualist, on the other hand, will be inclined to answer the original question negatively. To suppose that causes succeed rather than precede their effects in the region "after inversion" would imply for him that by manipulation of causes in that region one could "in principle" affect earlier events in it, and in just those circumstances where "prior to inversion" one could affect later events. He is unlikely to countenance the possibility of such wholesale actual alteration (rather than mere rewriting) of history. Consequently he would be expected to deny that the temporal direction of causation in the region "after inversion" is different from the original direction.

These are unlikely to be either the final or the most convincing replies of the physicalist and the conceptualist to our question. But they will serve to indicate the sort of connection that may be expected to obtain between a particular approach to causation and a view on the relation between temporal and causal asymmetry. I think that physicalist and conceptualist approaches contain elements of the truth about causation, but that neither is wholly correct. My goal in the present paper is to examine the relation between temporal and causal asymmetry in order to unearth clues which point the way to a fuller and more adequate understanding of causation.

## II

Since there has been much confused discussion of "the arrow of time" it is desirable to begin by isolating that use of this phrase which will be at the center of discussion in what follows. Glasses smash, cigarettes burn, rivers run down to the sea; people are born, grow old and die; useful energy is degraded into useless heat; light from the stars is dispersed into interstellar space until after exhausting their nuclear fuel they grow dark, and some may turn into black holes. Such processes are irreversible at least in the sense that they, but not their temporal inverses, occur in our region of the universe. They provide examples of a pervasive temporal asymmetry in the contents of our local region of spacetime. Any of these irreversible processes defines a direction of time in the following sense: certain distinctive stages of a process of such a type always precede and never succeed certain other distinctive stages of the same process (e.g. birth always precedes death).

It is important to notice that two asymmetric relations are involved here. The first is the temporal relation *being earlier than* (or its converse, *being later than*). The second is the physical relation, *being a distinctive stage of an irreversible process which in our local region of spacetime temporally precedes and does not succeed some other distinctive stage of that process*. While these two relations coincide in their application to distinctive stages of irreversible processes in our local region of spacetime, it is by no means obvious that they coincide elsewhere. Some views on the direction of time seek to define the former relation on the basis of the latter.[3] But the prospects of giving a convincing philosophical justification for such a definition look poor.[4] Certainly in this paper these two relations will be treated as distinct and independent; and it is the connection between the latter and the asymmetry of causation which will be the main focus of interest.

What are the relations between the various kinds of temporally asymmetric physical processes, and how pervasive are these processes? There is a strong temptation to seek a unified classification of all irreversible processes as those in which there is an increase in entropy, and to indicate their extent (if not to explain their occurrence) by appealing to the second law of thermodynamics. But this temptation will be resisted here since it is unclear how to state the second law in a form which is both true and broad enough to encompass processes like the dispersal of order through the emission of starlight, and the formation of black holes. It is preferable to avoid thermodynamic concepts like entropy by attempting to give an abstract account of the origin and subsequent course of an irreversible process in our local region of spacetime.

A typical irreversible process involves a single system consisting of many component parts. For example, a quantity of gas in a sealed container is made up of a vast number of molecules: and a normal pack consists of fifty-two playing cards. These component parts may each possess one out of a certain range of microscopic properties. A gas molecule may be in the left half of the container rather than in the right half; a card may be a diamond rather than one of the other three suits. If some microscopic properties are distributed among its component parts in certain ways, the system may itself consequently possess some macroscopic property. The gas may be confined to the left half of the container; the pack of cards may be arranged by suits. (Note that the terms 'microscopic' and 'macroscopic' as thus introduced need have no absolute significance, and may indeed have little to do with the spatial dimensions of a system or its parts.) Such a system undergoes an irreversible process as the microscopic properties of its parts develop and alter in such a way that the system acquires a set of macroscopic properties, or a macroscopic state, which is in some sense more disordered, less highly structured, or more to be expected than its initial macroscopic state. When a partition is removed, the gas which was previously confined by it to the left half of the container rapidly spreads out to fill the whole container uniformly; when the pack is shuffled for a while it ceases to be ordered in suits and acquires the characteristics of a "well-shuffled pack".

For such a process to be irreversible it is neither necessary nor sufficient that any physical laws governing the development of the microstate of the system (i.e. the set of microscopic properties of its component parts) be time-reversal invariant. All that is required for irreversibility is that the initial macrostate and its time-development be "typical" in the sense that most systems of this type developing in this way from the given macrostate would reach (and if undisturbed remain in) a macrostate indistinguishable from that reached by this system in the time concerned. Of course, as stated, this last requirement is quite vague. How can one give a precise characterization of the way the system develops? In what sense are the resulting macrostates indistinguishable? What exactly are the microstates and macrostates of a given system? More or less adequate answers to these questions can be given for specific types of process, but the details will vary from process to process, and for present purposes detail is best passed over in the interests of generality.

A characteristic of such an irreversible process is that at the start the system involved is in a macrostate of pronounced order or structure, but this is realized by an otherwise quite disorderly microstate. As the process unfolds, the development of the initial microscopic disorder "washes out" the initial macroscopic structure. Irreversibility is then the expected consequence of the special nature of the initial state, and so an explanation of irreversibility requires an account of why the system has such a special initial state.

An account of the origin of an irreversible process in our local region of spacetime is typically of one of two sorts. The special initial conditions are either the result of purposive human intervention, or they arise naturally as a concomitant of the physical interactions which produce the system in the first place. The occurrence of a pack of cards ordered by suits is explained by the first sort of account; the existence of a lake high in the mountains which serves as the source of a river is explained by the second sort of account. It will simplify the discussion considerably if for a while attention is restricted to naturally originating irreversible processes.

The origins of almost all naturally originating irreversible processes in the solar system are traceable ultimately to the single temporal asymmetry constituted by the fact that throughout geological time the sun has continued to emit vastly more radiant energy than it has absorbed. For example, the water in the mountain lake came from snow melted by the heat of the sun; this snow was deposited on the mountainside after atmospheric instabilities ("weather") largely due to the periodic heating of the atmosphere by the sun; and the water of the lake originally evaporated from an ocean after it was warmed by the sun. The origins of this temporally asymmetric behavior of the sun itself lie in whatever cosmogonic processes led to the gravitational clumping of a body composed predominantly of nuclear fuel of very light elements like hydrogen. There exist fairly plausible accounts of such cosmogonic processes, which permit the origins of most naturally originating irreversible processes in the solar system to be traced further and further back in cosmological time (with decreasing confidence in the correctness of the account!) towards the currently popular "big bang" origin of the universe.[5]

A system involved in a typical irreversible process did not always exist and will not exist forever. In Reichenbach's (1956) terminology, it 'branched off' from a wider system at some point, entering a state of (relative) isolation from its surroundings, and at some time it will very likely merge once again into a wider system. At the initial branch point, the system came into being with properties it inherited as an integral part of a wider system, as well as with properties resulting from the interaction which gave rise to it. Generally, the disordered microscopic state will be inherited, and the ordered macroscopic state will result from the originating interaction. But even where this is not so, a system involved in an irreversible process will typically have a special initial state, explicable in terms of the processes which led up to its formation, but there will be no corresponding explanation of its subsequent state in terms of the processes which "lead back to" that later state.

So much for temporal asymmetry; causal asymmetry may be dealt with more briefly. Causes produce and explain the occurrence of their effects; effects do not produce or explain the occurrence of their causes. The causal relation manifests several such general asymmetries, and these may be supplemented by reference to any number of instances in which the causal relation is clearly asymmetric. It is a requirement on any adequate analysis of causation that it explain (or at least explain away) the clear prima facie asymmetry of the causal relation, though this requirment is one that a number of analyses find surprisingly difficult to meet. In particular, both regularity and *sine qua non* analysts tend to put in this asymmetry almost as an afterthought by defining the earlier of two nonsimultaneous causally related events as the cause rather than the effect.[6] Such a definition threatens to trivialize the question to which this paper is addressed. But the threat may quickly be averted by pointing to obvious shortcomings of such a definition. First there are clear prima facie examples of simultaneous causation, moreover examples in which the causal relation retains its asymmetry. (Consider Kant's (1781) lead ball resting on a pillow and causing its depression; or gravitational attraction causing the downward acceleration of a stone.) Second, the possibility of backwards causation (an apparent prerequisite for precognition) cannot be ruled out by definition. Although causation is asymmetric, the precise connection between causal asymmetry and the asymmetry of the earlier/later relation must not follow as a trivial consequence of an adequate philosophical analysis of causation.

Just as some philosophical views seek to base a definition of the *earlier than* relation on temporal asymmetries in physical processes, so there are causal theories of time which seek to base a definition of the *earlier than* relation on the relation of causal asymmetry. If successful, such a causal theory of time would also threaten to tie the asymmetry of the *earlier than* relation to that of the causal relation. But weighty reasons may be added to those mentioned in the previous paragraph to render the prospects for such a causal theory of time rather dim.[7] The question as to the relation between causal asymmetry and temporal asymmetry remains open and interesting.

## III

This section examines three broadly physicalist approaches to causation and considers the difficulties each of them faces in giving an adequate account of the relation between causal and temporal asymmetry.

Consider first a strict regularity view of causation according to which a cause is a set of initial conditions for some true physical law governing the time development of a system, whose final conditions constitute the effect. I call this view strict because the regularities to which it appeals do not include statistical or approximate regularities like the second law of thermodynamics, but are to be found (if at all) near the frontiers of physical theory. The trouble with this view is that the only plausible candidates for exceptionless regularities governing the development of an isolated system in such a way that precise initial conditions uniquely imply particular final conditions are also such that precise final conditions uniquely imply particular initial conditions. To give just one example, according to Newton's laws of motion, the specification of final velocities and positions of a set of otherwise isolated point particles interacting via known forces will determine their initial values of velocity and position, just as a specification of the initial values of velocity and position will determine the final values of these quantities. On a strict regularity view there is nothing in the distinction between initial and final conditions to underlie the asymmetry of causation other than the mere temporal precedence of the former. So no purely physical feature has been shown to provide a basis for causal asymmetry.

A proponent of the strict regularity view may reply that temporal precedence *is* a purely physical relation; and that to provide a relation which is coextensive with causal priority (and would be even in counterfactual circumstances) *is* to provide a basis for causal asymmetry. In assessing this reply it is important to note the following feature of the view. If causation rests on exceptionless regularity plus temporal precedence, then a "temporal inversion" of the contents of a region of spacetime will not make causes succeed rather than precede their effects in that region; instead it will (somewhat arbitrarily) change causes into effects and effects into causes, while preserving the temporal precedence of causes. Appreciating why this interchange seems arbitrary will help one to see why the strict regularity view does not provide a basis for causal asymmetry.

A physicalist reduction of causation need not be offered as a conceptual analysis of our notion of causation. But if one knows to what physical relation causation has been reduced, it should become clear why we have the concept of causation that we do have. Now if causation reduces to a relation between a set of initial conditions in some exceptionless regularity and a particular final condition, then one may define an analogous relation of

$R$-causation as a relation between a set of final conditions in some exceptionless regularity and a particular initial condition. The strict regularity account does nothing to explain why we have a notion of causation but no notion whose physical correlate is $R$-causation. This suggests that one should look for a "thicker" physicalist account of causation which permits an appeal to details of the actual physical behavior of the world in explaining why our concept of causation is of a future-directed and not a past-directed relation.[8] If such an account were forthcoming it would also be clear why, had the world behaved differently (as, for example, in the temporal inversion considered in Section I), a past-directed notion of causation would have been more natural than our present future-directed notion.

Just such a "thicker" physicalist account seems to have been gaining support. In a recent paper David Fair argues that "physical science has discovered the nature of the causal relation for a large class of cases. As a first approximation, it is a physically-specifiable relation of energy-momentum flow from the objects comprising cause to those comprising effect"[9]. As the author points out later in his paper, one reason this can only be a first approximation is that even when energy-momentum flow is clearly involved in an instance of causation, it does not always flow in the right way! When the drop in temperature causes the pipes to crack by freezing the water they contain, energy flows from the pipes to the cooler surrounding air, not vice versa: very little energy or momentum in the president's finger would be required to initiate a nuclear holocaust, and there is no reason to expect any of it to flow to Moscow to contribute to the destruction of sections of that city. But let us suppose that the second or third approximation is able plausibly to deal with such counterexamples.

How are causal and temporal asymmetry related according to this account of causation? Fair claims it as a virtue of his account that it provides an immediate explanation of causal asymmetry by transference from the asymmetry of the earlier/later relation implicit in the analysis of causation as energy flow from cause to effect. But he also wishes to retain backwards causation at least as a conceptual possibility, which could be realized if it were "possible for a positive amount of energy to flow from an object at a time to an object at a time earlier, as measured in the same reference frame" (p. 241). This raises several difficulties for the account.

First, it may be doubted whether it is a conceptual possibility for a positive amount of energy to "flow backward in time" from an object $A$ to an object $B$. If this is not a conceptual possibility then neither, on this account, is backwards causation; and yet there are reasons to believe that whether or not it occurs, backwards causation is at least a conceptual possibility. But let us provisionally grant that energy may "flow backward in time" from $A$ to $B$. How would this differ from energy flowing "forward in time" (i.e., flowing!) from $B$ to $A$? It is important for this view that such a distinction can be made; for if it cannot, then perfectly innocent causal processes may be counted as instances of backwards causation. For example, if we let $A$ be the cold outside air and $B$ the water in the pipes, then why should we not conclude that the cracking pipes provide an instance of backwards causation?

Perhaps the example of tachyonic processes (Fair, p. 241) will show how positive energy could "flow backward in time" in such a way as to constitute a case of backwards causation? If there were such things, tachyons would be particles which travelled faster than light relative to any inertial frame in Minkowski spacetime (the spacetime of special relativity). Consequently, a positive energy tachyon emitted at event $E$ and absorbed at $E'$ (with $E$ earlier than $E'$ in frame $F$) would appear in some frame $F'$ to be absorbed before it was emitted, provided $E'$ is earlier than $E$ in $F'$. This appears to make sense of the idea of energy "flowing backward in time" (relative to $F'$) from the later cause $E$ to the earlier effect $E'$. Now there has been considerable dispute as to whether $E$ or $E'$ should be considered the cause in such a case, relative to $F'$.[10] But if (relative to $F'$)$E$ is regarded as the cause and $E'$ the effect, so that we have a potential case of backwards causation, then we also have a counterexample to Fair's analysis (though perhaps not to some refinement of his first approximation). For since a tachyon with positive energy relative to $F$ has negative energy relative to $F'$, the direction of (positive) energy flow will be from effect to cause, not from cause to effect!

I conclude that it is far from clear whether on energy-momentum flow accounts, backwards causation is really a possibility. But does this constitute a criticism of such accounts? Even if one assumes that our notion of causation leaves open the conceptual possibility of backwards causation, it does not seem to follow that a physicalist reduction of this notion must retain that possibility. Consider an analogous reduction. While our *concept* of sound does not exclude sound transmission in outer space (recall that battles in *Star Wars* tended to be noisy!), in fact, given that sound is just the propagation of longitudinal vibrations through a medium, sound couldn't possibly be transmitted through outer space. However, there are important differences between this example and proposed physicalist reductions of causation, and these differences defeat the intended import of the example.

One who offers a physicalist account of causation is not in a position to say precisely what physical features correspond to each instance of causation. For example, even if causation did correspond to energy-momentum flow, one would not have to hand an actual reduction of the causal relation, in so far as physics has not reached a final classification of cases of energy-momentum flow. Proposed physicalist accounts of causation are mere sketches for actual reductions of the causal relation, which are intended to be filled in as and when physics succeeds in discovering exactly what each element in the sketch corresponds to. Moreover, as physics develops, new and perhaps radically different kinds of causation may be recognized. No physicalist reduction-sketch can be adequate if it rules out this possibility. In particular, in so far as backwards causation is a conceptual possibility, no physicalist reduction-sketch can be adequate if it rules out the possibility that some future physical theory might involve backwards causation. So, unlike the case of sound, a physicalist account of causation *must* allow the possibility of (the development of a physical theory which involves) backwards causation. This kind of possibility may be called 'potential physical possibility', and contrasted both with present physical possibility and with ultimate physical possibility. As long as it is a conceptual possibility, whether or not it is physically possible according to present theories or even according to a hypothetical ultimate physics, backwards causation is potentially physically possible, and no correct physicalist reduction-sketch can deny this. (In addition to the energy-momentum flow account, the strict regularity account may also be criticized on the grounds that, without radical modification, it too denies the physical possibility of backwards causation.)

Nevertheless, the energy/momentum flow approach suggests another physicalist account of causation which does promise to make room for backwards causation. For one of the most significant general features of naturally occurring irreversible processes is precisely that they involve an energy/momentum flow. This suggests that a physicalist account of causation be somehow based on properties of these irreversible processes.

One of the most provocative attempts to base an account of causation on properties of irreversible processes is that of Hans Reichenbach (1956, Chapter 4). Reichenbach offers the following definition of the causal relation: "the cause is the interaction at the lower end of the branch run through by an isolated system which displays order: and the state of order is the effect" (p. 151). In this definition, by 'the lower end of a branch' is meant the end at which the isolated system's (macro)entropy is less: no physical law entails that this must also be the earlier end of the branch. Later he makes it clear that this definition is intended only to capture the core notion of causation, and that by extension the term has come to have a wider meaning (p. 156). For Reichenbach, the paradigm example of a cause is the interaction in which a branch system originates; and of an effect, the macroscopically ordered initial state of the branch system.

Now it is a contingent, though very important, fact about our local region of spacetime (and indeed, according to currently popular cosmologies, about the whole of spacetime after the first few minutes of cosmic expansion) that branch systems are almost always formed in ordered initial states, whose order then dissipates as the systems evolve toward disordered final states. But suppose that some branch system behaves differently: suppose that its state immediately prior to the interaction in which it merges into some wider system is macroscopically ordered, so that it is of such a kind as to permit easy and reliable predictions of its earlier states, leading back to a disordered initial state. And suppose further that this initial state is such that readily accessible knowledge of it would not have permitted easy, reliable prediction of the system's subsequent development. Then on Reichenbach's definition it follows that the interaction in which this branch system merged to form part of a wider system is one in which cause succeeds effect!

Here is an example which is extremely unlikely to occur, but whose occurrence would be compatible with all known physical laws. Consider a sample $S$ of a gas created in a microstate $M$ which may be defined in the following way. Let $S'$ be a sample of gas initially enclosed in the left half of a container by a partition; and suppose that the partition has been instantaneously removed, allowing the (subsequently isolated) gas to spread into the whole container. Suppose that at a later time when $S'$ has reached a state of macroscopic equilibrium, the microscopic state of $S'$ and its container is $M'$. Then $M$ is defined as follows: the positions of all the molecules of $S$ and the container in state $M$ are the same as those of all the molecules of $S'$ and the container in state $M'$; and the velocities of all the molecules in state $M$ are equal in magnitude but opposite in direction to those of the molecules in state $M'$. A sample of gas created in state $M$ would develop so that its initial state "unravels" until at a particular instant all its molecules are present in the left half of the container; and we can suppose that at that precise instant a partition is inserted, interrupting the state of isolation of the gas. On Reichenbach's definition, the insertion of the partition caused the gas to have adopted its final macroscopically ordered state! And yet what really caused the gas to adopt its final macroscopic state was in fact the natural development of the gas following whatever fortuitous circumstances caused it to be in the remarkable initial state $M$.

A natural modification of Reichenbach's account provides a way of avoiding the conclusion that in certain anomalous structure-producing processes cause succeeds effect.[11] It is to take the direction of causal asymmetry in any one irreversible process to correspond not to the direction of increasing disorder in that process but rather to the direction of increasing disorder in almost all irreversible processes. This modification avoids the above problem, and finally yields a plausible physicalist account at least of many cases of causal asymmetry which conforms to the expectations of Section I. That is to say, on this last account, a "temporal inversion" of our region of spacetime will result in causes succeeding instead of preceding their effects.

But now this modified version of Reichenbach's account appears once again to have eliminated the potential physical possibility of backwards causation in our local region of spacetime (and of forwards causation in its temporal inversion). For a consequence of the modification is that either for all irreversible processes throughout our local region cause precedes effect, or for all irreversible processes throughout our local region, cause succeeds effect. A new approach is required which, while not immediately excluding the possibility of a class of examples of backwards causation in our local region of spacetime, will not automatically treat all anomalous structure-producing processes as falling within this class. Such an approach will be developed in the next section.

## IV

The metaphor of the causal net will provide the key ingredient for the proposed account of the relation between causal and temporal asymmetry. Cause/effect pairs do not occur in isolation from one another, but are linked together. The effect of one cause will generally be the cause, or a part of the cause, of further effects: and causes typically are not themselves uncaused. Any determination as to which of a pair of causally related events caused the other must preserve consistency with the results of other such determinations. If *B* and *C* are causally inseparable (token) events, in that they are causally related and an event of neither type occurs without an event of the other type, a determination of which caused which will generally follow from the investigation of their other causes and effects: for example, if *B* has causes other than *C*, but *C* has no causes other than *B*, then *B* caused *C*.

Even for the strict physicalist, the absence of any non-relational physical properties of two causally related events sufficing to pick out one as the cause of the other need not leave their causal asymmetry open. For the events may be uniquely locatable within a causal net containing other event pairs whose physical features do suffice to establish their causal asymmetry; and the causal asymmetry of the original pair will then be derivable from their location in the causal net.

We can now see how to discriminate between an anomalous structure-producing irreversible process and a genuine case of backwards causation. In the former case, the apparently disordered initial state will turn out to be causally produced by earlier events, though if these can in turn be traced to no coordinating preceding cause the account of its causal production will appear very unsatisfactory as an explanation. But in the latter case, it will turn out that there is no account of the causal production of the initial state by earlier events. An example may help here.

If a stone is dropped into a still pool, ripples will spread out uniformly from its point of entry and gradually die away. Suppose that the reverse process were to occur: a still pond gradually develops incoming circular ripples which grow as they converge upon a point, until as the last ripples reach that point a stone pops out of the pond and decelerates upward, leaving the surface of the pond quite smooth. This reverse process need not violate any physical laws: but of course it is nevertheless not to be expected. Would such a reverse process constitute at least a prima facie instance of backwards causation (the stone's subsequent emergence from the pond causing the ripples to have developed)? The answer to this question depends on how the sequence of events could be fitted into the causal net. If it could be established that each individual molecule of water in the pond was caused to move in the way it did by preceding molecular collisions, then the case could be considered an incredibly unlikely example of a normal, future-directed causal process. But if it could be determined that the motion of each individual molecule of water had no preceding sufficient cause, then we should have a satisfactory explanation of the event only if it involved a case of backwards causation (though one may still deny that this constitutes *evidence* for backwards causation).

Assignment of causal asymmetry is like solving a set of simultaneous equations: the assignment of causal asymmetry to one causally related event pair in a region depends on the assignment to all other causally related event pairs in that region. Pursuing the analogy, the existence and uniqueness of a solution to a set of simultaneous equations depends on whether the set determines, underdetermines, or overdetermines the values of its unknowns. What is to prevent the assignment of causal asymmetry to event pairs in a spatiotemporal region from being underdetermined with respect to an overall interchange of causes and effects? A final version of Reichenbach's account answers this question by reference to the nature of the temporal asymmetry of almost all irreversible processes in a region. This version does not hold that the causal asymmetry of each irreversible process is tied to its temporal asymmetry (which would count atypical irreversible processes as cases of backwards causation); nor does it tie the causal asymmetry of each causally related event pair to the temporal asymmetry of almost all irreversible processes in the region (thus failing to allow for the possibility of backwards causation). Instead this version requires the overall order of causation in the causal net to be determined as that corresponding to the temporal asymmetry of almost any irreversible process in the region.

V

Is the physicalist approach to causal asymmetry presented in the previous section satisfactory? I think not. There is another candidate for the job of assigning overall direction of causal asymmetry to the causal net: it is the criterion that of two causally related events *A* and *B*, if a person can produce an event of type *A* without producing an event of type *B* but not vice versa, then *B* causes *A*.[12] The conceptualist will argue that this criterion is fundamental to causal asymmetry while the criterion of correspondence to the direction of most irreversible processes in a region is not. In any potential conflict between these criteria (as there seems to be in the case of the "temporally inverted" region of spacetime), the latter criterion must bow to the former: consequently the physicalist approach to causation must be wrong. The direction of causal asymmetry is indexical with respect to our causal powers.[13] But the physicalist is not left without a rebuttal to this conceptualist argument, and the ensuing debate promises to lead to a deeper appreciation of the nature of causation.

An initial physicalist response focuses on the notion of "our causal powers" required by the conceptualist's criterion for overall causal asymmetry of the causal net. The response is to assert that since our possession of any causal powers is contingent upon the existence of just that pervasive local temporal asymmetry on which the physicalist's rival criterion for overall causal asymmetry is based, the conceptualist's criterion turns out to be parasitic upon rather than ascendant over that of the physicalist. This response has two stages which will be examined in turn.

What reason is there to believe that our possession of any causal powers is contingent upon the existence of a pervasive local temporal asymmetry? Recall from Section II that in addition to naturally originating irreversible processes, very many local temporally asymmetric processes are initiated by human intervention, as someone prepares the system involved in the requisite special initial state. Now there is some plausibility to the suggestion that human preparation of such an initial state is made possible by the temporal asymmetry inherent in naturally originating irreversible processes. Consider, for example, the preparation of an initial state of a gas in which it is confined to the left half of its container, but where the microstate is otherwise quite disordered. One way to prepare this initial macrostate would be to evacuate the other half of the container with a pump. But the successful operation of the pump depends upon the occurrence of certain irreversible processes whose initial microstates have not been prepared by human intervention: and that the state of the remaining gas is microscopically disordered is a result, not of direct human intervention, but ultimately of some naturally originating irreversible process.

Now it may reasonably be objected that while humans perhaps could not prepare the initial state of an irreversible process without the aid of other naturally originating irreversible processes, nevertheless the exercise of human causal powers is not restricted to the preparation of such states. But there is a stronger reason to believe that a close link exists between naturally originating irreversible processes and the exercise of human causal powers. For a person would be literally unable to lift a finger were it not for the occurrence of naturally originating irreversible processes required for the conversion of chemical energy stored in the muscles into kinetic energy of bodily motion. So it is quite plausible to suppose that, creatures of the flesh as we are, our possession of any causal powers is indeed contingent upon the existence of a pervasive local temporal asymmetry.

But would this show that the conceptualist's criterion for overall causal asymmetry of the causal net is secondary to a physicalist criterion based upon the existence of pervasive temporal asymmetries in the spatiotemporal region in question? Since application of each of the two criteria yields the same result in our local spatiotemporal region, in order to assess their priority relations one must consider circumstances involving other spatiotemporal regions, either distant or counterfactual. Suppose first that as a result of astronomical investigations it becomes scientifically well established that in some spatiotemporally distant region of our universe types of process characteristic of the pervasive temporal asymmetries of our local region are reversed. (To the extent that according to our best current science this is not the case, the supposition may be expected to be contrary to the facts.)

If the physicalist criterion works by comparing the prevailing direction of temporal asymmetry in this spatiotemporal region to that in ours, then the conclusion will be that since this direction is reversed, so is the direction of causal asymmetry in that region. For example, the occurrence of nuclear reactions in a star will count as the cause of the earlier absorption of radiant energy by that star, and so will explain the otherwise inexplicable convergence of electromagnetic radiation onto that body. However, this seems clearly incorrect. For our attainment of any knowledge of such a spatiotemporally distant region must be an effect of causes within that region. Hence, as the conceptualist will point out, the causal net within our region is joined to the net within this other region; and this linkage suffices to impose a direction of causal asymmetry on the distant region by transference from our local causal asymmetries. Now it may be that when this transference is made the direction of causal asymmetry in the distant region turns out to coincide with that derived from the physicalist's simple comparison of the prevailing directions of temporal asymmetry in the two regions. But just because this would be a coincidence, the above application of the physicalist criterion leads to results only contingently coincident with the prior assignment of the direction of causal asymmetry within the distant region, which is inherited from that of our local spatiotemporal region via the causal links between the two.

It is interesting to note in passing that there is a problem, in reconciling primitive conceptualist intuitions to the possibility of backwards causation, which is opened up by acceptance of the priority of this last method of assigning the direction of causal asymmetry. For, as pointed out in Section I, the conceptualist may find it hard to countenance the possibility of backwards causation. But it may be that projection from our causal powers into some distant region via the links of the causal net reveals the existence of causally connected events $E$ and $E'$ in that region such that although $E'$ occurs before $E$, we can produce an event of type $E'$ only by producing an event of type $E$ and not vice versa. Nevertheless, in that case the resulting tension among conceptualist intuitions is best resolved by admitting that $E$ causes $E'$. For this is the only way to maintain the conceptualist principle that the independently manipulable event is the cause, not the effect. Note also that if the region containing $E$ and $E'$ is in the distant future, then one would not be bringing about the past if one produced $E'$ by producing $E$.

Now it may be held that though we can have no direct causally mediated knowledge of the existence of a spatiotemporally distant region in which the direction of temporal asymmetries is reversed, still there may be general scientific arguments to the effect that the existence of such regions is very likely.[14] The conceptualist criterion would be inapplicable to any such regions, since no causal link exists between them and our local region: but the physicalist criterion could still be applied (though only hypothetically if we had no reason favoring one direction of temporal asymmetry in one of these regions over the other). What are we to say of the application of the physicalist criterion in such a case? Note that to apply the conceptualist criterion it is not required that we presently have knowledge of processes within the distant region via causal interactions. All that is required is that some event, perhaps in our distant past or future, is causally linked both to our region and to the distant region.[15] If we assume that the conceptualist criterion is inapplicable because there is no causal link whatever between us and the region, even via events in our distant future or past, then the case seems parallel to that of a region in some merely possible world; and though the physicalist criterion can be applied, there seems little reason to accept the results of its application. Rather one might take this as further evidence supporting the conceptualist criterion: for in a case like this where it proves inapplicable, we have serious doubts about the sense of speaking of a direction of overall causal asymmetry.

The physicalist may hold that his criterion cannot be applied region by region, but must be applied to the universe as a whole. But then a potential (though unlikely) outcome of astronomical investigations could be the discovery that the temporal asymmetries in our local spatiotemporal region constitute a cosmic anomaly. In that case we might be forced to admit that the local direction of causal asymmetry is the reverse of what we had thought; really the explosion causes the spark, not vice versa. This constitutes a *reductio ad absurdum* of the revised application of the physicalist criterion.

A more plausible physicalist position would be to maintain that while the direction of causal asymmetry throughout the universe is settled by its local direction and the links provided by the causal net, nevertheless the local direction of causal asymmetry is determined by the direction of locally pervasive temporal asymmetries. Now if all the physicalist view comes to is the assertion that these directions in fact coincide, then the view is platitudinous and not in conflict with the conceptualist view. An interesting physicalist view extends this coincidence to counterfactual situations so as to assert that in any possible universe, local temporal asymmetries are related to the local direction of causal asymmetry in the way the two are related in our universe: this is the sense in which actual local temporal asymmetries determine the actual local direction of causal asymmetry.

What notion of possibility is being appealed to here? It cannot just be (potential) physical possibility, for there are physically possible worlds without any temporal asymmetry, and other physically possible worlds for which the term 'local' has no sense since there are no people in those worlds by reference to whose spatiotemporal location it may be provided with application. The set of admissible worlds must be restricted to those populated by agents sufficiently like us, with pervasive temporal asymmetries in the spatiotemporal region of these agents. Clearly this set is rather ill-defined, and this represents a first objection to the latest physicalist view. But apart from this problem in clearly stating the view, how is the physicalist to argue for it?

Recall that the key premise in the physicalist's argument for the priority of a physicalist criterion for the direction of overall causal asymmetry was that our possession of any causal powers is contingent upon the existence of a pervasive local temporal asymmetry. The content of this premise is presumably that in any possible circumstances in which we can intelligibly describe ourselves as having causal powers, there will exist a pervasive local temporal asymmetry, which permits us to have these powers.

To further clarify the content of the premise we must once more ask what counts as a possible set of circumstances. The considerations advanced in favor of the premise did nothing to show that if we were disembodied Cartesian souls our exercise of causal powers would be conditional in this way, so if the premise is to be supported by the considerations advanced in its favor, the possibility of our being disembodied Cartesian souls must be excluded. I suggest that a set of circumstances be counted as possible only if

(1) it is physically possible,

(2) it contains a physical realization of objects that are appropriately characterizable as agents, and

(3) these agents may reasonably be considered to be persons.

If the physicalist premise is understood in this way, the following objection is forestalled. We can consistently describe a world in which we are disembodied spirits with causal powers; and in such a world the overall direction of causal asymmetry is fixed not by the prevailing temporal asymmetries but by the "causal direction" in which these causal powers are exercised. Hence the physicalist criterion could fail in such a world. The objection is forestalled because such worlds are irrelevant to the assessment of the physicalist claim that our possession of causal powers is contingent upon the existence of a pervasive local temporal asymmetry.

Unfortunately, though the physicalist's premise may thus be rendered defensible, it does not lead to the desired conclusion, which was to be that the conceptualist criterion for assigning overall causal asymmetry is actually subordinate to a physicalist criterion. The most that the physicalist's premise can be said to establish is that the applicability of the conceptualist criterion is contingent upon the existence of a pervasive local temporal asymmetry. It does not follow that there are circumstances in which the conceptualist criterion is applicable, but where its application leads to results which are incorrect, and moreover incompatible with the correct results flowing from the application of the physicalist criterion. It does not even follow that there are circumstances in which the conceptualist criterion is inapplicable, but the physicalist criterion may be correctly applied. (It may be useful to recall here the earlier discussion of causally disconnected regions.) Hence the physicalist has failed to show that the conceptualist criterion for the assignment of overall causal asymmetry is subordinate to his own. The direction of causal asymmetry is indexical to our causal powers, though an interesting set of physical facts underlies both.[13]

At this point one may wonder why it is important for the physicalist to argue for the *precedence* of some physicalist criterion over a conceptualist criterion, rather than for their mere coincidence. Now recall the counterfactual situation with which the opposition between physicalist and conceptualist was introduced. The physicalist intuition which motivated this opposition was that one may simply read off, from an objective physical description of some portion of spacetime containing no human beings (or other agents), which events cause which other events. This intuition seems to be what lies behind the physicalist's attempt to show that his criterion for assignment of overall causal asymmetry must not simply agree with, but must take precedence over the conceptualist's criterion. But is this intuition essential to a distinctively physicalist view?

One might argue that a physicalist account of causation which *denied* that one may simply read off from a physical description of some portion of spacetime – containing no human beings (or other agents) – which events cause which other events, had after all failed to locate causation in the physical world. The need to introduce talk of agents into the description appears to make the notion of causation undesirably anthropocentric. I will argue that the physicalist may be able to avoid these difficulties: but first it is necessary to examine this notion of anthropocentricity, and to relate it to a more precise statement of the claim that the direction of causal asymmetry is indexical with respect to our causal powers.

In his study of indexicals, David Kaplan has distinguished between what he calls the *context of use* of a term occurring in a declarative sentence, and the *circumstances of evaluation* of the truth of that sentence.[16] A term is *indexical* if its referent (or extension) varies systematically with its context of use. By claiming that the direction of causation is indexical to our causal powers, I intend rather that the truth-conditions of '*A* is causally prior to *B*' are dependent in a certain way on the circumstances of evaluation of that sentence. Specifically, assessment of the truth-conditions of '*A* is causally prior to *B*' (and hence also of '*A* causes *B*') requires that the assessor locate her/himself in the causal net containing *A* and *B*. (Note that since I am not offering a definition of causal priority or of causation, no vicious circularity is involved here.) It follows that '*A* causes *B*' has no absolute truth-value independent of the relations between *A*, *B* and a person *e* (*the evaluator*) who assesses the truth-value. Compare 'The store is across the road' whose truth-value varies according to the spatial location of the evaluator.

This last example shows that an expression may be such that the truth-values of certain sentences containing it are sensitive to their circumstances of evaluation, even though the expression is not in any interesting sense anthropocentric. But if the features of the circumstances of evaluation to which these truth-conditions are sensitive are features of the evaluator *qua human being* rather than just of his/her physical situation, then there is a sense in which the relevant expression may justifiably be called 'anthropocentric'. For example, 'being more useful than' may be considered anthropocentric in this sense, because whether *A* is more useful than *B* depends (apart from *A* and *B*) on the *aims* (and abilities) as well as e.g., the spatial location of the evaluator. Now the sense in which the notion of causation seems to have an anthropocentric component is this: the truth-value of '*A* causes *B*' is sensitive to features of the evaluator, qua human being. Only the sentence '*A* causes *B*, according to *e*' has truth-conditions independent of the (actual) circumstances of evaluation, and in this sentence the reference to a human being has become explicit.

However, there is still a way open for the physicalist who wishes to reject the conclusion that the notion of causation has an anthropocentric component. If he could display the physical basis *Q* of *e*'s causal powers, then explicit reference to a human being might be eliminated. That is to say, the sentence '*A* causes *B*, according to *e*' could be replaced by the sentence '*A* causes *B*, relative to *Q*' with the same truth-conditions in all possible circumstances (in the sense of 'possible circumstances' delineated above). And then '*A* causes *B*', though still sensitive to circumstances of evaluation, would not display the sort of sensitivity which would justify the conclusion that, on the physicalist's account, causation has an anthropocentric component.

Suppose that this is the way the physicalist proceeds. If he is successful then he may arrive at a physical basis for the causal relation, even though an essential part of the specification of this basis will consist in a description of physical features of an agent. It will not be true that the physicalist can simply read off from a physical description of some portion of spacetime containing no agents, which events cause which other events; but it will be true that only a canonical *physical* description of a person *e* will be required for the physicalist to be able (in principle) to say whether or not *A* caused *B*, according to *e*. Such a physicalist account would yield a criterion which in no way conflicts with the conceptualist criterion, and therefore is not ascendant over it. But coincidence with a conceptualist view counts as a *virtue* for this type of physicalism.

Such a physicalist project is of an altogether different magnitude from the limited project of finding a physical basis for the truth of '*A* causes *B*' solely in the events *A* and *B* and simple physical events in their immediate spatio-temporal region. Pending significant progress toward its completion, we are justified in maintaining the view that *our* notion of causation has an anthropocentric component which no simple physicalism can eliminate. I shall conclude by considering some objections which may be raised against this last view.

## VI

One objection against the view that our notion of causation has an ineliminable anthropocentric component goes as follows. If the truth of '*A* is causally prior to *B*' really depends on who evaluates it, then it should be possible for *e* and *f* to evaluate the truth of the sentence oppositely. But this never happens, and indeed it *could* never happen.

In reply to this objection, I accept that we do in fact agree on our attributions of causal priority, and that this agreement is not contingent. But the basis for this agreement lies not in any objective relation between events like *A* and *B*, but rather in the causal relations that obtain between members of our epistemic community. All members of my epistemic community are agents with whom I can interact causally. Moreover, we all exercise our causal powers in the same "causal direction", for otherwise we would not interact interact in such a way as to form a mutually responsive, intercommunicating, epistemic community.[17] It may or may not turn out to be a conceptual truth that anything I can recognize as an agent exercises its causal powers in the same "causal direction" as I do. If so, then all attributions of causal priority *must* coincide. If not, then there could be, perhaps in some distant galaxy, an agent whose causal powers were exercised in the opposite "causal direction" to mine.[18] I suspect that such an agent could not be part of my epistemic community: at any rate, (s)he does not form part of it. That is why *we* all agree on our attributions of causal priority.

Another objection is that on this view, if there were no agents, there would be no direction of causal asymmetry: and that is an absurd consequence of the view.

Now it is not clear to me just how absurd a consequence this is. Indeed, progress towards a physicalist reduction of causal asymmetry along the lines considered toward the end of the last section would, I think, make it very natural to deny the existence of a direction of causal asymmetry in a world with no physical states or processes of type $Q$. But, whether or not one admits that it is strictly speaking incorrect to attribute an overall direction of causal asymmetry to a world without agents, it is still very natural to do so. And on the present view of causal asymmetry one can understand why this is so.

It is indeed a consequence of the view that evaluation of the truth-conditions of '$A$ causes $B$' requires that the evaluator be related to $A$, and/or to $B$. In the central case, this relation will itself be causal. If there are no agents in some "possible world", then no evaluator is causally related to events in that world. But still, there may be a very natural way of "projecting" an evaluator in this world into that possible world. Suppose, for example, that the human race had never evolved on Earth: would the stars still have caused light to be emitted into outer space? One is justified in answering 'Yes' in so far as there is a natural way of "projecting" an evaluator into this "possible world" so that in that world the evaluator's causal powers would be exercised in the same "causal direction" as they are in this world. The actual evaluator bears a relation of "projection + indirect causal connection" to the stars in this possible world, and though this is not a purely causal relation, its obtaining makes it at least very natural to say that stars would have emitted (rather than absorbed) radiation even if the human race had never evolved on Earth. Compare this with the question 'Do cars in New Zealand drive on the left?'. Although 'to the left of' is sensitive to the spatial orientation and position of the evaluator, there is a natural way of "projecting" the evaluator into the intended circumstances of evaluation (facing along a road in New Zealand in an upright position) which makes the answer

'Yes' a natural one, whatever the *actual* present spatial relation between the evaluator and New Zealand.

Finally one might raise the following difficulty. There is an inconsistency between the view that the notion of causation has an anthropocentric component, the view that physics is an important part of the search for causes, and the view that physics progresses by eschewing anthropocentric notions.

The simple resolution of this difficulty is to note that, for all that has been established in this paper, there may be nothing anthropocentric in the symmetric relation of causal connectedness. Physics could deliver explicit statements about the (undirected) causal connections in the causal net, without violating a ban on anthropocentric notions by referring to the direction of causal asymmetry: and by relating ourselves to the causal net we could use the characterization physics gives of it to derive knowledge of causes. But a deeper reply would be that physics does not issue in explicit statements of causal connections, but rather, characterizes the causal net in *non-causal* terms. We arrive at knowledge of causal connections by interpreting such characterizations in causal terms and at the same time relating ourselves to this causal net so as to impose a causal asymmetry on it. Hence physics itself may describe the world without using any causal terminology, while yet yielding knowledge of causes. This reply is deeper since it would suffice even if the symmetric relation of causal connectedness were shown to be importantly anthropocentric, a possibility I have not considered here. Furthermore, and most importantly, this reply would also be consistent with the notable absence of causal terms from the vocabulary of much of physical theory.[19] It is interesting to note that this reply will only prove satisfactory, however, if a weaker physicalist thesis is true. In order for it to be possible for us to relate ourselves to the characterization, in physical terms, of the causal net, it must be possible for us to characterize *ourselves* in these same physical terms, at least to a limited extent.

## NOTES

[1] Bertrand Russell (1912), in a famous paper, expressed this negative conclusion as follows: "the reason physics has ceased to look for causes is that, in fact, there are no such things. The law of causality, I believe, like much that passes muster among philosophers, is a relic of a bygone age, surviving, like the monarchy, only because it is erroneously supposed to do no harm."

[2] Though the counterfactual situation has not been very precisely described, a more precise, though more verbose, description could be given for Newtonian spacetime and

(with respect to a given inertial frame) for Minkowski spacetime. One might doubt whether such a situation can consistently be described in certain relativistic spacetimes. This paper does not address the question as to the relation between causal and temporal asymmetry in any spacetime for which no such consistent description can be given. I doubt that this constitutes an important limitation.

[3] Cf. Reichenbach (1956), Grünbaum (1963).

[4] Cf. Earman (1974). But see also Sklar (1981).

[5] See, for example, Davies (1974).

[6] See Earman (1976).

[7] See, for example, Sklar (1974), Chapter 4.

[8] By using the term 'future-directed' I do not intend to claim that our concept of causation excludes the possibility of isolated cases of backwards causation. Indeed I shall soon appeal to this possibility myself.

[9] Fair (1979), p. 220. See also Quine (1973), Section 2; and Byerly (1979).

[10] See Earman (1972) and Maund (1979) and the literature cited therein.

[11] In fact this may well be Reichenbach's own considered view, despite his explicit definition of cause quoted earlier. Certainly, this latter view would square better with a causal theory of time order, which Reichenbach also proposed in this book.

[12] In this paper, I intend by 'criterion' not 'way of telling that', but rather 'condition whose obtaining makes it the case that'.

[13] Though suggestive, like all epigrams, this will turn out to involve misleading oversimplifications (see below, pp. 97–8).

[14] Boltzmann, for example, once put forward the view that our local spatial region deviates from cosmic equilibrium by being of anomalously low but increasing entropy: on such a view, in the distant past (during its formation as a cosmic anomaly) temporal asymmetries in our local spatial region must presumably have been reversed, even though we have no knowledge of this.

[15] Thus, for example, event pairs within the event-horizon of a black hole, or event pairs in the history of objects outside a particle horizon, could be ordered with respect to the direction of causal asymmetry, according to this conceptualist criterion.

[16] Kaplan (1977).

[17] Note that it does not follow that backwards causation is impossible: merely that if there are cases of backwards causation, they do not figure importantly in the mechanisms of ordinary human and social interaction.

[18] An intelligible description of such an agent would require the employment of a reversed *temporal* description of his/her actions. This would not imply that "time can flow backwards", but it would involve the admission that the relation of temporal precedence shares the anthropocentricity of the relation of causal priority. Though I have not explicity addressed this issue in the present paper, I should perhaps make it clear that I am sympathetic to the suggestion that temporal priority, as well as causal priority, is an anthropocentric notion: cf. Healey (1981).

[19] This absence was noted by Russell (1912). But cf. Earman (1976).

## BIBLIOGRAPHY

Byerly, H.: 1979, 'Substantial Causes and Nomic Determination', *Philosophy of Science* 46, 57–81.

Davies, P. C. W.: 1974, *The Physics of Time Asymmetry*, Surrey University Press, London.

Earman, J.: 1972, 'Implications of Causal Propagation Outside the Null Cone', *Australasian Journal of Philosophy* **50**, 222–37.

Earman, J.: 1974, 'An Attempt to Add a Little Direction to "The Problem of the Direction of Time" ', *Philosophy of Science* **41**, 15–47.

Earman, J.: 1976, 'Causation: a Matter of Life and Death', *Journal of Philosophy* **73**, 5–25.

Fair, D.: 1979, 'Causation and the Flow of Energy', *Erkenntnis* **14**, 219–50.

Grünbaum, A.: 1963, *Philosophical Problems of Space and Time*, Knopf, New York.

Healey, R.: 1981, 'Statistical Theories, Quantum Mechanics, and the Directedness of Time', in *Reduction, Time and Reality*, ed. by R. Healey, Cambridge University Press, Cambridge.

Kant, I.: 1781, *Critique of Pure Reason*.

Kaplan, D.: 1977, 'Demonstratives' (unpublished manuscript).

Maund, G.: 1979, 'Tachyons and Causal Paradoxes', *Foundations of Physics* **9**, 557–74.

Quine, W. V. O.: 1973, *The Roots of Reference*, Open Court, LaSalle, Illinois.

Reichenbach, H.: 1956, *The Direction of Time*, University of California Press, Berkeley.

Russell, B.: 1912, 'On the Notion of Cause', in *Mysticism and Logic*, George Allen and Unwin, London.

Sklar, L.: 1974, *Space, Time and Spacetime*, University of California Press, Berkeley.

Sklar, L.: 1981, 'Up and Down, Left and Right, Past and Future', *Noûs* **15**, 111–29.

W. H. NEWTON-SMITH

# TEMPORAL AND CAUSAL ASYMMETRY

## 1. CONCEPTUALISM AND PHYSICALISM

Healey has sought to improve our understanding of causation through a critical evaluation of two rival approaches, that of the physicalist and that of the conceptualist, with particular reference to the question of the direction of causation. The physicalist, as characterized by Healey, looks to our best physical theories to see what, if anything, in the physical world corresponds to causation. If nothing does he concludes that causal talk is illegitimate (Healey, p. 79). For Healey's conceptualist no discovery in physics could have such exciting implication for the question of the status of causal talk. The conceptualist's starting point is rather a hypothesis in philosophical anthropology to the effect that our causal concepts had their origin in the primitive human experience of producing changes in objects. That, I would have thought, is uncontentious. The more interesting claim is that while we have modified our primitive ancestor's concept of causality, it retains to this day such a close connection with human agency that "causes are potential means by which humans could, at least 'in principle', bring about their effects" (Healey, p. 79). This anthropocentric aspect of causality means that "the direction of causal asymmetry is indexical with respect to our causal powers" (Healey, p. 98).

By sharpening a substantial issue concerning causation Healey has improved our understanding of it and his arguments display the deficiencies in particular physicalist approaches (i.e., Fair and Reichenbach). However, as will be argued in this section of this paper, he has not made out the case for the anthropocentric conceptualism which he favours. Nor has Healey provided a satisfactory characterization of what it is for an account to be anthropocentric or to be physicalistic. In the second section of the paper I argue through a clarification of these notions that Healey has misunderstood the nature of the physicalist programme and that, in fact, on a proper understanding of that programme Healey's own arguments lend support to it. There are non-anthropocentric accounts of the direction of causation which Healey has not considered. One of these due to Mackie will be examined in the third and final section of the paper. Mackie's account will not do as



*R. Swinburne (ed.), Space, Time and Causality*, 105–121.

it stands. However, suitably modified it is plausible. Within the confines of the present paper I cannot fully develop a case for such an account. However, a brief consideration will help to clarify the physicalist approach.

Healey assumes for the sake of argument an unanalysed directed temporal relation and an unanalysed undirected relation of causal relatedness. I will follow him in this in order to focus specifically on his notion of directed causation. Given that $A$ and $B$ are causally related events his 'criterion' for the direction of causation is: "if a person can produce an event of type $A$ without producing an event of type $B$ but not *vice versa*, then $B$ causes $A$" (Healey, p. 92). This criterion is not a verification condition. It is not a "way of telling that" but rather a "condition whose obtaining makes it the case that" (Healey, Note 12). I will call this *Healey's condition* or *criterion* and since it involves direction I will refer to it as holding in the above case from $B$ to $A$. Clearly Healey's condition does not give correct answers to questions about the direction of causation between all pairs of events if applied *directly* to those events. For example, suppose that the Big Bang cosmological model is correct and that it successfully predicts the gross features of the contemporary universe by reference to the Big Bang. Under these assumptions there is no doubt that the Big Bang was a cause of the current state of the universe and not vice versa, even though on no reasonable construal of what a person can do in principle is it the case that a person could bring about a state of the type of the current state of the universe by bringing about a state of the type of the Big Bang. And Healey would agree, I presume, that no account of causality which precluded us from holding this would be acceptable.

It might seem that Healey's invocation of the metaphor of the causal net meets the above objection. However, a consideration of this reveals deeper difficulties in Healey's account of causation. In displaying this we will assume with Healey that the extension of the relation of causal relatedness is fixed on the set of all events and that that set is ordered in time. Let $E_1$ and $E_2$ be events in our local region of spacetime which are such that on Healey's condition, $E_1$ causes $E_2$. Let $E_{-m}$ and $E_{-n}$ be distant events which are such that Healey's condition is not satisfied with regard to them (i.e., they occur at times in the past when the preconditions for the existence of agents like ourselves were not satisfied). Part of the causal web could be represented as follows with '–' representing causal relatedness and '→' representing directed causality.

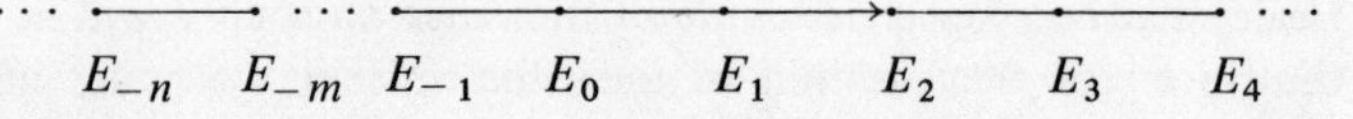

Can we extend the relation of directed causality to distant events in the net by reference to the particular case of directed causality between $E_1$ and $E_2$? Given that with Healey we do not wish to rule out the possibility of backwards causation we cannot do so simply by reference to the temporal ordering of events and the relation of causal relatedness. Indeed, on Healey's picture, the extension is determined by such principles as "if *B* has causes other than *C*, but *C* has no causes other than *B*, then *B* caused *C*". (Healey, p. 90). Let us suppose that on such principles it turns out that $E_{-n}$ causes $E_{-m}$, even though Healey's condition is not satisfied from $E_{-n}$ to $E_{-m}$. At this juncture the pressing question is the nature of the connection between a claim of the form "Event *B* caused event *A*' and 'Healey's condition is satisfied from *B* to *A*". Either the former claim entails the latter claim or it does not. Presumably Healey does not intend the connection to be one of entailment. For in that case we would be committed to inconsistently holding both that Healey's condition is not satisfied from $E_{-n}$ to $E_{-m}$ and that it is satisfied from $E_{-n}$ to $E_{-m}$. And if we construe the connection as being contingent the example shows that it is not an invariable regularity. There will be no shortage of events in our universe between which directed causality obtains and in regard to which Healey's condition is not satisfied. Thus it is not even as a matter of contingent fact a necessary condition of directed causality.

Healey's condition is more plausible if interpreted as a sufficient but not necessary condition of directed causation. Whether Healey intends to be understood as asserting that the satisfaction of his condition from *B* to *A* entails that *B* causes *A* or that as a matter of empirical fact if it is satisfied from *B* to *A*, *B* causes *A* does not affect the following argument. Given that Healey's condition is a sufficient but not necessary condition of directed causation, there may be other sufficient conditions. For instance, as Healey concedes, his refined version of Reichenbach's physicalist criterion (Healey, p. 92) applied to our region of spacetime provides the same results as his criterion. Why then should the physicalist criterion not be recognized as providing another sufficient condition of directed causation? In which case there will be worlds devoid of agents in which there will be a direction to causation even though Healey's criterion is not applicable (but see below pp. 110ff for a discussion of Healey's attempt to extend his criterion to agentless worlds via his notion of projection). If it is conceded to the physicalist that his criterion provides another sufficient condition of directed causality, Healey's exciting and substantial claim that directed causality is anthropocentric lapses. He has at best reminded us that there are anthropocentric conditions the satisfaction of which constitute cases of directed causation. But if there are non-anthropocentric conditions the obtaining of which makes it true that there is a direction of causation, causation is not in any interesting sense anthropocentric.

If anything of Healey's anthropocentricity thesis is to survive it is crucial that he show that his criterion take priority over the refined physicalist criterion. But since "creatures of the flesh we are our possession of any causal powers is indeed contingent upon the existence of a pervasive local temporal asymmetry" (Healey, p. 93), the alleged priority will have to be established, as Healey notes, through a consideration of distant or counterfactural spacetime regions. Before considering such cases we need to make precise the notion of priority. Healey denies that he is providing either an analysis or a definition of directed causation. Presumably, then, we are to employ our unanalysed intuitive notion of directed causation and compare how well the results of his criterion and the results of the physicalist criterion fit with judgments made using our unanalysed notion. To establish the priority of his criterion, he will have to display conditions in which his gives the correct answer and the physicalist gives the wrong answer or conditions in which his is applicable but the physicalist's is not.

Healey first considers a possible spatiotemporally distant region to which we have no causal links and in which the direction of temporal asymmetries is reversed relative to our region of spacetime. As Healey notes, the conceptualist criterion is not applicable to such a region but the physicalist criterion is. But, Healey claims:

> there seems little reason to accept the results of its application. Rather one might take this as further evidence supporting the conceptualist criterion: for in a case like this where it proves inapplicable, we have serious doubts about *the sense of speaking* of a direction of overall causal asymmetry. (Healey, p. 95) (my italics).

But this simply begs the question. For unless one had already assumed what is at issue, namely the priority of the conceptualist criterion over the physicalist criterion, one could argue as follows: Given the perfect correlation between the satisfaction of these two criteria in the region of spacetime where we can consider the result of using each, we should make the judgment of causal direction on the basis of the criterion which is applicable in regions in which only it is applicable. If Healey had shown that there is something semantically deviant in talking about directed causation in such a case where the physicalist criterion would license such talk he would have established his priority claim. The phrase 'the sense of speaking' suggests that this is his line of reasoning. However, not only has he provided no argument for this claim, it is difficult to see how he could, given that his is not trying to define or analyse the notion of directed causation. It should be noted that whether or not there are reversed temporal asymmetries in the causally isolated region is not relevant. Healey's criterion is inapplicable in either case due to the causal isolation, and the physicalist's is applicable in both cases.

Healey agrees that the applicability of the conceptualist criterion is contingent upon the existence of a pervasive local temporal asymmetry which could be used in applying the physicalist criterion (Healey, p. 97). He claims that this alone does not show that there are situations in which the conceptualist criterion is inapplicable but in which the physicalist criterion gives the correct answer. While it does follow the physicalist can argue that there are such situations (including the one considered above) and that defeats Healey's claims for the priority of his criterion. To reiterate, the physicalist's argument runs as follows. The two criteria match up in our local region of spacetime. In this region the results of either fit with our intuitive judgements as to the direction of causality. There are possible regions in which the physicalist but not the conceptualist criterion is applicable. It is entirely reasonable and part of standard scientific practice to apply the criterion which is applicable. That being so Healey has not shown that his criterion has priority.

The scientific practice referred to above can be illustrated by reference to judgments about spatial distance. We have a criterion (in Healey's sense of the term) for such judgments based on the transportation of rigid rods. We also have a criterion based on sending light signals. The former which is more closely connected with our intuitive notion of spatial distance gives the same results as the latter in situations in which both are applicable. In cases where relativistic complications mean that the former is inapplicable, we certainly do not refrain from making judgments. We employ the latter criterion. In a similar vein the physicalist will argue that judgments as to the direction of causality should be made by reference to his criterion in situations in which it but not the conceptualist's is applicable.

To sustain his anthropocentricity thesis Healey has to establish the priority of his criterion over that of the physicalist through a consideration of distant or counter-factual spacetime regions. Possible worlds without agents provide another test case for the priority claim. For Healey's criterion as initially stated is not applicable to any such world whereas the physicalists may be. On first considering a world without agents Healey claims that it is at least very natural to deny that there would be an overall direction of causal asymmetry and, further, that conceptualism can explain why this is natural. Imagine a world as like this as possible save for the fact that due to some unhappy accident no life evolved. I would have thought that the most natural thing to say is that there would be a particular overall direction of causation. A physicalist can explain this for the conditions he takes as constituting the existence of an overall direction obtain. If on Healey's account of causation there would be no direction, that is a sufficient reason for rejecting the account.

In any event Healey has sought to extend his account with a view to licensing talk of directed causation in some agentless worlds. It is to be noted that any extension of the account which does not undermine the anthropocentric thesis will have to make essential reference to agency in fixing the direction in agentless worlds. On the extended account we can talk of an overall direction if we can 'project' an evaluator into the possible world in a 'natural way', "so that in that world the evaluator's causal powers would be exercised in the 'same direction' as they are in this world" (Healey, p. 100). To begin with let us restrict consideration to a possible world such as that noted above which is sufficiently like the actual world that the physical conditions do not preclude the possibility of an agent's existing. Let $E_1$ and $E_2$ be types of events which occur in the actual world, which are such that an agent can bring about an instance of type $E_2$ by bringing about an instance of type $E_1$ and not *vice versa*. Let us suppose further that on the physicalist criterion as well as on Healey's the direction of causation in the actual world is from instances of $E_1$ to those of $E_2$. Instances of $E_1$ and $E_2$ occur and are causally related in the possible world. What would ground a claim that it was natural to project an agent into the possible world so that the direction of his causal powers was from $E_1$ to $E_2$ and not *vice versa*? One might seek to ground this by appeal to a counter-factual to the effect that if there were an agent like ourselves in that world he could bring about $E_2$ by bringing about $E_1$ but not *vice versa*. But what in turn could ground that counter-factual? The physicalist has an answer: if on his physicalist criterion $E_1$ causes $E_2$ and if $E_1$ is a type of event an agent like ourselves can produce, he could bring about $E_2$ by bringing about $E_1$. But if this grounds the counter-factual, the extended conceptualist criterion is parasitic on the physicalist criterion, contrary to Healey's priority claim.

Another possibility would be to count as natural the projection which preserves as far as possible the direction of causal processes in the actual world. But then the anthropocentric element in the extended account drops out. For Healey as he himself notes cannot establish the priority of his criterion by reference to the actual world. And so to select an overall direction in a possible world by reference to directions in the actual world, does not show the direction in the possible world to be dependent on agents unless it does so in the actrual world but that is precisely the question at issue. To reiterate the problem: Healey's criterion is implausible if it precludes talk of directed causality in the sort of agentless possible world noted above. Healey has sought to meet this objection by extending his account to include the 'projection' in a 'natural' way of agents into such worlds. However, it is not clear that one can explicate the notion of naturalness without either making the extneded criterion parasitic on the physicalist or making the extended criterion non-anthropocentric.

Even if Healey's extended criterion can be deployed to meet the objection in relation to the sort of world considered above there are other worlds in which it cannot. For instance, consider a world consisting mostly of point sources distributed throughout space from which light fronts frequently emerge and in which the physical conditions preclude the formation of objects complex enough to constitute agents. Given that the light fronts emerge from the point sources but never contract onto them, we would be justified in thinking of that world as having an overall direction of causal asymmetry. If we posit the radiation of the point source as causing the expanding wave front, we have an explanation of the cordinated motion of the photons in the wave front. If we were to take the causality to operate in the other direction, it would be totally mysterious what produced the cordinated motion. Hence, we should posit the direction of causality as being from the radiation of the source to the expansion of the wave front. Given that the vast majority of processes in this possible world are of this character and are temporally irreversible we would be justified in asserting an overall direction to causality. This familiar style argument reminds us that we have ways of reasoning about causal processes that can lead to the justifiable ascription to a direction without invoking Healey's anthropocentric criterion. Since as I argue below even Healey's extended criterion does not give a direction to causation in the possible world under consideration, it has to be rejected.

There is a problem about understanding what it would mean to 'project' an evaluator into a possible world in which agents could not exist. For Healey's anthropocentric thesis involves the claim that the truth-conditions of judgments of causal direction are sensitive to features of the agent *qua* human being (Healey, p. 98). If the anthropocentric thesis is to be preserved, the direction, if any, of causal processes in agentless worlds must depend on causal powers that humans would have in such worlds. But given that humans could not exist the notion of their causal powers in that world is not well-defined. Healey's extended criterion fails to be applicable in some cases where there would be an overall direction of causal asymmetry.

## 2. ANTHROPOCENTRICITY

According to Healey and others [1] our notion of directed causation has an anthropocentric component. Mackie [2] and others deny this. However, it is not entirely clear what is at stake. Consequently an attempt will be made in this section to clarify the notion of anthropocentricity. One result of which will be to show that rather than support an interesting anthropocentricity thesis, Healey's arguments tend to support the opposite conclusion. The anthropocentricity arises for Healey from the alleged fact that "the direction of causal asymmetry is indexical with respect to our causal powers" (Healey, p. 97). The indexical element in turn arises from the fact that "the truth-conditions of '*A* is causally prior to *B*' are dependent in a certain way on the circumstances of evaluation of that sentence" (Healey, p. 98). Healey notes that a sentence such as 'the store is across the road' is in this sense indexical without being in any interesting sense anthropocentric. An indexical sentence will be anthropocentric if "these truth-conditions are sensitive to features of the evaluator *qua human being* rather than just of his/her physical situation" (Healey, p. 98). It will be convenient to follow Healey in distinguishing these two claims.

Healey denies that he is analysing or defining the notion of directed causation. In view of that the claim of indexicality is somewhat surprising. For indexicality is a semantical feature and as such could be established only through a semantical analysis. Further, even if one took the conceptualist as offering an account of the meaning of '*A* causes *B*' it does not follow from Healey's version of conceptualism that '*A* causes *B*' is indexical. If the conceptualist holds that *A* causes *B* means that I or something I can recognize as an agent can bring about *B* by bringing about *A* but not *vice versa* and if it is a conceptual truth (as Healey suggests it might be) that anything I can recognize as an agent must exercise its causal powers in the same 'causal direction' as I do, all attributions of causal direction must coincide. (Healey, p. 100). In this case assertions of causal direction are not sensitive to circumstances of evaluation.

Even if we set aside the question as to whether a non-semantic account of causation can support a semantical conclusion about causation, Healey has failed to show that causal direction is indexical. If '*A* causes *B*' is indexical, its truth-conditions are dependent on the circumstances of evaluation. This means that there must be possible circumstances in which the sentence has one truth-value and other circumstances in which it has the other truth-value (where the referent of '*A*' and '*B*' are the same in both cases). Given his criterion for the direction of causality, he cannot be agnostic on the question as to whether it is a conceptual truth that anything I can recognize as an agent exercises its causal powers in the same direction as I do. If it is, the truth value of '*A* causes *B*' will not vary with circumstances of evaluation. To show indexicality, Healey has to describe a situation in which causal powers are exercised in the opposite direction. He suggests that this would involve an agent the intelligible description of which would require reversed temporal descriptions of his/her actions (Healey, Note 18). That is, for example, the agent's death comes before its birth. But given that Healey does not rule out the conceptual possibility of backwards causation, it might be more acceptable to describe such agents as indulging in backwards causation. In which case they could recognize our causal processes as occurring in the reverse temporal order than theirs and assign the same direction to the order of causation in our region as we do. That is, they recognize their causation as operating from later to earlier times, our causation as operating in the opposite direction and hence our causation as operation in the direction we recognize it as operating. Obviously the discussion of such agents needs a development that I am unable to provide here. However, unless Healey provides a detailed description of circumstances relative to which our correct judgment that '*A* causes *B*' is correctly said to be false, he has not shown that judgments of causality are indexical. Healey in his discussion has sought to explain why there is general agreement on the attribution of causal direction among members of our epistemic community given that causal direction is indexical. However, there is a simpler explanation of the agreement: causal direction is not indexical.

Healey's claims that the truth-conditions of sentences ascribing directed causality are anthropocentric in the sense that these are sensitive to features of the evaluator *qua* human being. This claim remains. For it could be true even though causality is not indexical. For instance, the sentence 'Icabod is in pain' is sensitive in this way without being indexical. This example suggests one a possible interpretation of the notion of anthropocentricity: a predicate is anthropocentric if it is a term for a property the instantiation of which is *mind-dependent*. '*x* is in pain' and '*x* likes *y*' would be clear examples of such predicates which I will call *mind-dependent* or *MD* predicates.

Healey's doubts about the sense of applying the notion of directed causation in worlds devoid of agents suggests that he has this in mind. However, as I have argued, he has failed to show that is is illegitimate to talk of directed causation in worlds devoid of humans. And, if we accept his extended analysis involving the notion of projection, directed causation fails to be mind-dependent even on his own account. For that extension seems intended to legitimatize talk of directed causation in agentless worlds.

Sometimes the term 'anthropocentric' is used to refer to predicates in the analysis of which *MD* terms enter essentially. I will call such predicates *MAD* for '*mind analytic dependence*'. For example, if '*x* is red' means '*x* would look to normal humans in standard situations the same colour as this pillar box', 'red' is *MAD*. The subjunctive element in such analysis means that a *MAD* predicate may still apply in conditions in which there are no humans. Hence, *MAD* predicates are not necessarily *MD* predicates. A conceptualist who offered a Healey-like account of the meaning of causation, would hold that causation is anthropocentric in this sense. Healey's criterion for '*B* causes *A*' is 'a person can produce an event of type *A* without producing an event of type *B* but not *vice versa*', and *MD* predicates occur essentially in it. If his criterion were intended as a meaning analysis of '*A* causes *B*', causation would be anthropocentric in the *MAD* sense. However Healey is not offering an meaning-analysis, but an account of the conditions that make ascriptions of causal direction true. This suggests a related but weaker sense in which causation has an anthropocentric aspect: a sentence of the form '*A* causes *B*' is true, only if a sentence into which *MD* predicates occur essentially is true. A predicate meeting this condition will be said to be *MAD**. If this is the sense in which Healey holds that causation is anthropocentric it is too generous a sense to give an interesting thesis. For I suspect that all theoretical predicates of physics are *MAD**. All observations terms are *MAD* since an explication of their meaning must make reference to human experience. The meaning of all theoretical terms while not exhaustively explicable by reference to observation terms are nontheless partially determined via connections with observations. Therefore, for any theoretical sentence to be true some sentence involving observation terms will have to be true (which as in the case of Healey's criterion may be subjunctive or counter-factual). Hence all theoretical terms are *MAD**.

There is another possible interpretation which will be developed through the consideration of the following fable. There is a tribe, the Herns, who use the term 'red' as given in the schema above. They also use a term 'fred' in a similar fashion. That is, there is a paradigm fred and general agreement on which objects look fred in standard conditions. Neither term is obviously mind-dependent. For the Herns apply these terms to objects and think that unperceived objects are or are not red or fred while unperceived. A wandering physicist comes on the scene. Neither term occurs in his theories. It contains instead terms like 'wave length', 'Ånstrom units' etc. The physicist has theories that explain everything in the physical world and hence, not surprisingly, he has come to think that the only really real properties are properties for which there are terms in his theories or in a definitional extension of his theories. He accepts entirely the claim of the Herns about what they mean by 'red' and 'fred'. Not having these terms in his theories, he looks to see if there is a physical correlate for these terms. That is, a property for which he has a term and which is instantiated if and only if the term in question is held to be instantiated by the Herns. Our physicist discovers that 'red' has a physical correlate and 'fred' does not. He concludes that redness is a really real property. For as he understands this notion, really real properties are those that are instantiated in the world and these are properties designated by terms in his theory. The term 'red' as used by the Herns picks out a property picked out by a term in his theory, even though the physicist and the Herns have different means of detecting instantiations of that property. On the other hand, for our physicist, fredness is not a really real property. Fredness for the Herns is like goodness for us: it lies in the eyes of the beholder.

Our physicist can give some hard content to his metaporical claim that fredness lies in the eyes of the beholder. If he considers a possible world in which the same laws hold as in his world and which is as similar as possible except for the absence of conscious agents, given a description of the world, in terms of the predicates which occur in his theory, he can say whether there are any red objects in that world. He cannot say whether any objects are fred. *Ex hypothesi* there would be no way in which if grounding counterfactuals about how objects in that world would look if there were human agents in it.

This little story suggests another interpretation of the notion of an anthropocentric predicate. To explicate this we define a *respectable physical property* to be a property designated by a term occurring in a correct physical theory (including in the theory any definitional extension of it). A predicate with a physical correlate is a predicate which *designates* a physical property. Anthropomorphic predicates on this fourth understanding of that notion are those which lack physical correlates. I will call these *MP* predicates for *'mind-projected'*. At this stage I am only concerned to characterize a plausible version of physicalism. At the end of the paper I will examine the legitimacy of the physicalist's assumption that predicates without physical correlates do not express real physical properties. The story suggests an explication of physicalism which would make the physicalist position not only immune from Healey's criticism, but actually supported by his own arguments. The physicist in my story understands really real properties to be those expressible in his theory and that is not an unreasonable understanding. For why, he would argue, should we assume that there are more properties instantiated in the worrld than are referred to in a successful total theory of nature? Given this orientation, when faced with a predicate which is used systematically, he looks to the best current theories to see if there is a physical correlate. If there is not, he will not proscribe the use of that predicate but will take a negative view of the ontological significance of its use.

The physicalist notices a fairly systematic use of a notion of directed causality in our local region of spacetime which he can accept is correctly described by Healey's account. He looks for a physical correlate with a view to seeing if anything in the world corresponds to this talk. His early attempts to find a correlate are defeated by Healey's arguments. Happily for him, Healey comes up with a refined characterisation of a correlation which it is reasonable to assume holds. Further, Healey's arguments tend to show that this is a law-like rather than accidental correlation. For its holding is a precondition of the applicability of Healey's own criterion. However, Healey denies that this shows a physicalist criterion based on the local temporal asymmetries is *superior* to the conceptualist criterion. But on my understanding of physicalism he need not advance this claim in the first place. He can accept Healey's *MAD* account and ask still if there is anything in the world corresponding to directed causation. That question of the anthropocentric character of directed causality in the *MP* sense remains whether or not Healey's account correctly describes our use of a notion of directed causality. On this understanding of physicalism, the physicalist is one who takes seriously the question as to what, if anything, corresponds to our talk of directed causality. If the answer is negative he will regard causation as being anthropocentric in the sense articulated. I also claimed, provocatively, that Healey has advanced the physicalist case. For he has accepted that the existence of pervasive local temporal asymmetries is an empirical precondition for the operation of his conceptualist criterion. Healey further accepts that if the physicalist fixes the local causal direction by reference to these asymmetries and then extends the causal ordering via the causal net it is plausible to suppose that the resulting order matches that which would be produced by his manipulation criterion. Thus he is committed to holding that it is plausible to suppose that there is a physical correlate for directed causation (albeit a complex relational correlate), and, thus, that in one sense causality is not anthropocentric.

In the first section of the paper it was argued that Healey's criterion does not have priority over the physicalist criterion. In this section a number of ways of construing the notion of an anthropocentric predicate have been considered. It has been shown that on his own account, causation is not anthropocentric in the *MD* or *MAD* sense and that while it may be anthropocentric in the *MAD** this is uninteresting since it shares this feature with all theoretical predicates. On the final sense, a predicate is anthropocentric if it lacks a physical correlate. But Healey's arguments support the conclusion that directed causality is not anthropocentric in this sense. For he has argued that in our region of spacetime there is a non-accidental correlation between his criterion and the physicalists. In other words, there is a physical correlate of directed causation, at least in this world. I have not represented the physicalist as conceding that there are possible worlds with directed causation in which his criterion fails. But he could concede this and still maintain that causality has a physical basis in the actual world and in this my physicalist differs greatly from Healey's.

An analogue will help to reveal his reasoning. The physical basis of temperature in the actual world is molecular motion. There is a law-like connection between a certain kind of substance being warm to the touch and its having a certain degree of molecular motion. It is, however, possible to envisage worlds in which objects varied in temperature (as determined by touch by agents like ourselves) in which some different mechanism produced this phenomenological effect. That molecular motion would not be the physical basis of temperature in those worlds does not mean that it is not in this world. Similarly, the physicalist, using Healey's refined version of Reichenbach's criterion, can argue that even if there are worlds in which that condition is not the physical basis of directed causation, it is so in this world. The best conclusion given our current physical theories is that directed causation has a physical basis in this world and that it is not in any sense an anthropocentric notion. Furthermore, there is a good sense in which the physicalist criterion is the more basic. The temporal asymmetries in the world constitute for the physicalist the basis of the direction of causation. Given a direction of causation, he can aspire to explain what manipulations by agents like ourselves can produce. For the conceptualist, the manipulations by agents like ourselves can constitute the basis of the direction of causation.

It seems bizarre to suppose that from this as the basis of causation one could explain the occurrence of the temporal asymmetries in the world. That is, it is reasonable to hope we might explain our causal powers by reference to the existence of temporal asymmetries. But it is unreasonable to think we could explain the existence of temporal asymmetries by reference to our causal powers. Hence, a criterion of causation tied to temporal asymmetries is more basic from the explanatory point of view than one tied to human causal powers.

## 3. NON-ANTHROPOMORPHIC CAUSATION

In this section of the paper I develop a non-anthropomorphic analysis of the notion of directed causality. My starting point is Mackie's discussion of the direction of causation in which he objected (Mackie, p. 181)[3] to the 'deplorable anthropocentricity' of von Wright's interventionalist analysis of causation, an account not essentially different from Healey's. Von Wright, does, however, offer it as a semantic analysis. Mackie endeavours to fix the direction of causality using a notion of fixity. For Mackie an event is fixed at time $t$ if it has occurred at $t$ or if there is then some sufficient condition for its later occurrence. $A$ is the cause of $B$ if $A$ and $B$ are causally connected in direct line and if there is no time at which $B$ is unfixed and $A$ is fixed (or, in other words, whenever $A$ is fixed, $B$ is fixed).

This analysis has many attractive features. First, it is non-anthropocentric. Secondly, as Mackie points out, it gives a criterion that matches up with an interventionalist or conceptualist criterion where both are applicable. Thirdly, the analysis does not rule out the possibility of backwards causality. A form of backwards causation, hereafter cited as *BC*, considered by Mackie is the following: $A$ and $B$ are causally related events. $A$ occurs at $t_2$, $B$ occurs at $t_3$. At $t_1$ $A$ is not fixed (there is then no sufficient condition for $A$'s later occurrence). At $t_1$ $B$ is fixed (there is then a sufficient condition of the later occurrence of $B$). $B$ is the cause of $A$ on Mackie's analysis for any time at which $A$ is fixed, $B$ is fixed. $A$ is not the cause of $B$ for there is a time, namely $t_1$, at which $B$ is fixed but $A$ is not.

In spite of these attractive features the analysis is implausible. For it entails that the notion of directed causation has no application to a strictly deterministic universe. In such a world, at any time, every event is then fixed either in virtue of having happened or in virtue of there being then a sufficient condition for its occurrence. To quote Mackie, "If you have too much causation, it destroys one of its own most characteristic features"

(Mackie, p. 91). This is most implausible. If we came to believe that we live in a deterministic world, we would continue to think that (except in cases of *BC*) that earlier states could be directed causes of latter states. We would reveal this belief, in among other things, explaining the current state of the universe by reference to the Big Bang and not *vice versa*. Either this would be sheer prejudice on our part or there is something wrong with Mackie's analysis. I incline to the latter view and will offer an alternative analysis of the direction of causation that shares the merits of Mackie over Healey (i.e., not anthropocentric), without necessarily depriving casuality of a direction in a deterministic world.

To this end I offer a different definition of fixity, show that it permits backwards causation and directed causation in a deterministic world and then consider the rationale for this particular definition. If our definition of fixity is to allow for the possibility of directed causation in a strictly deterministic universe, it must avoid making all events in such a world fixed at all times. If it is to allow for the possibility of *BC*, it cannot simply equate fixity with having occurred. To allow both these possibilities, we define the notion of being *fully fixed* as follows: At time $t$ event $A$ is fully fixed if *both* $A$ has occurred and either there is no sufficient condition for $A$ or a sufficient condition for $A$'s occurrence has occurred. Using Mackie's pattern of analysis, $A$ is the cause of $B$ if and only if $A$ and $B$ are causally inseparable and any time at which $B$ is fully fixed is a time at which $A$ is fully fixed.

Consider a fully deterministic world. Let $A$ occur at $t_1$ and $B$ occur at $t_2$ where $A$ and $B$ are causally inseparable. In view of the assumption of determinism, there is at time $t_1$ a sufficient condition for the occurrence of $B$. But $B$ is not fully fixed until time $t_2$. $A$ is fully fixed at time $t_1$. Thus any time at which $B$ is fully fixed, $A$ is fully fixed. Thus $A$ is the cause of $B$. Therefore, there is directed causation in a deterministic world.

Consider a *BC* situation of the type envisaged by Mackie. At time $t_1$ there is a sufficient condition for the occurrence of $A$ at time $t_3$ but no sufficient condition for the occurrence of $B$ at time $t_2$. At $t_3$ $A$ is fully fixed. $B$ is not fully fixed until after $t_2$ for it is not fully fixed until after the occurrence of a sufficient condition for its occurrence. Consequently, any time at which $B$ is fixed is a time at which $A$ is fixed, thus $A$ is the cause of $B$. Therefore, backwards causation is not ruled out on this account of directed causation.

This alternative definition of fixity could be defended simply on the grounds that it provides an analysis of causation which gives the correct answers. It allows for the possibility of *BC* without precluding the possibility of directed causation in a deterministic world. But perhaps something more can be done to provide a rationale for saying that events are not fully fixed until after the occurrence of sufficient condition (if there is one). In the case of events that are not subject to backwards causation, being fully fixed amounts to having occurred. That seems reasonable. Even if a sufficient condition obtains, the event still has to occur. An event *E*, which is backwardly caused, is not fully fixed until its cause has occurred. And that too seems reasonable. For when *E* occurs, it lacks something. Its cause is yet to occur. There is something about it that is not yet fixed. Something has happened but we have to wait for its *raison d'etre* to be. My notion of being fully fixed captures the sense in which *BC* events are strangely incomplete – their explanation turns on something yet to occur.

On Mackie's analysis there are possible worlds without directed causality, for instance, any strictly deterministic world. And on Healey's accounts some worlds will lack directed causality. A law-governed world in which the physical conditions prevent the existence of beings with causal powers would lack directed causality. And on certain views of agency, Healey's analysis would, like Mackie's, mean that there was no directed causality in a strictly deterministic world. On general empiricist grounds I would be most dubious of my modified version of Mackie's if every world with causally inseparable events was a world with directed causality. However, any world in which time is closed is a world without directed causality as will be shown. To obtain a picture of closed time it is fruitful to run through the following imaginary thought scenario. Imagine we live in a world which could be quite like this one and that we believe that time is linear, non-ending and non-beginning. Suppose we come to have evidence that a cyclical cosmological model best fits the world and that determinism is true. Further, the evidence supports the conjecture that the universe has always and will always be running rhough a qualitatively identical sequence of states. Given an inclination to a relationalist theory of time and a belief in the identity of indiscernibles, it would be natural to hypothesize that corresponding times in the apparently distinct cycles were really identical. This then gives the hypothesis that the structure of time is that of a closed curve.

If we attempt to apply our notions of past, present and future to such a world we find that at the time *t* of the occurrence of event *E, E* is then not only present but also both past and future. In addition every event is before and after every other event. This means that if we were to apply the analysis of causation given above, any pair of causally inseparable events is such that the directed causal relation holds both ways between them! For any event is fixed at any time. At any time you like, any event you like has occurred and also a sufficient condition for its occurrence has also occurred. In point of fact we cannot adequately characterise temporal order in such a world using our standard notions of past, present and future or before and after. We need instead to use the complex relation of pair-separation as articulated by van Fraassen (or corresponding tense-operators). Given then that our standard, common-place temporal notions of past, present and future or before and after do not apply, the notion of directed causality does not apply for it is defined by reference to these notions. Thus a closed time world is a world in which there would be no directed causality.[4]

If the modified version of Mackie's account is accepted as an analysis of our notion of directed causation, we have still, with the physicalist to ask what, if anything, expressible in our phsyical theories corresponds to our talk of directed causation. I suggested that the refined version of the physicalist criterion provides the answer. We can now see that even if that should be false and that there is no physical correlate of directed causation, it is not legitimate for the physicalist to conclude that directed causation is a suspect notion which we impose on a world that knows it not. For given that *A* and *B* are causally connected the truth-conditions for '*A* causes *B*' provided by the analysis obtain or do not independently of ourselves even if the relation of directed causation has, mysteriously, no physical correlate. There may after all be more things going on in the world than is dreamt by the physicalist. However, as things stand we have good reason to think that our non-anthropocentric notion of directed causation does in fact correspond to something expressible in terms of our current physical theories.

## NOTES

1 See for example G. -H. von wright, *Explanation and Understanding* Routledge and Kegan Paul, (London, 1971), Ch. II.

2 J. L. Mackie, *The Cement of the Universe*, Oxford University Press, (Oxford, 1974), Ch. 7.

3 Pagination refers to Mackie, *op. cit.* Mackie's analysis has been criticised on the grounds that it precludes directed causality in a deterministic world and Mackie himself has conceded that the analysis will not do as it stands.

My modification is designed to meet this particular objection. Even if further modifications may be necessary a consideration of the analysis as it stands is useful in clarifying an aspect of the physicalist's position. (See especially pages 24–5.)

4 For a further discussion of closed time see my *The Structure* of Time Routledge and Kegan Paul, (London, 1980), Ch. III.

CAUSALITY AND QUANTUM MECHANICS

NANCY CARTWRIGHT

# HOW THE MEASUREMENT PROBLEM IS AN ARTIFACT OF THE MATHEMATICS*

## 0. INTRODUCTION

One of the most troubling assumptions of quantum mechanics is the projection postulate. This is the postulate which causes many to believe that there is a special role for consciousness in nature. The projection postulate describes what happens not in any circumstances but in those special circumstances in which an observation is made; and what is supposed to happen when an observation is made is very different from what happens in other interactions. When an observation is made, a 'reduction of the wave packet' occurs which does not occur otherwise.

The way I look at it, the projection postulate is responsible for the notorious 'measurement problem' in quantum mechanics. Nature behaves differently when a measurement is made from how it behaves otherwise. But what constitutes a measurement? How does nature known when to behave one way and when another? A lot of people try to solve this problem by giving up the projection postulate. I don't think we can give it up. But we should realize that it has little to do with measurement. Reductions of the wave packet occur other times as well.

Still, it seems we are left with a critical characterization problem. When does nature obey the projection postulate and when does it behave in the more ordinary way, in accord with the Schroedinger equation? In fact, this is no real problem. Once we have divorced reduction of the wave packet from measurement, the conceptual problems surrounding the projection postulate disappear. I am going to argue that the characterization problem is a pseaudo-problem which arises from taking the mathematics too literally. The measurement problem turns out to be an artifact of the mathematics.

## I. WHY WE SHOULD BELIEVE IN THE PROJECTION POSTULATE

Reduction of the wave packet can best be explained by an example. We know that electrons have both a particle and a wave aspect. The wave aspect can be seen in diffraction experiments. A beam of electrons is sent through a diffraction grating and then falls onto a recording screen. If each of the



*R. Swinburne (ed.), Space, Time and Causality*, 125–134.

electrons behaved like a rock, passing through either one hole or another of the gaps in the grating, we would expect to see a particular kind of pattern on the recording screen. But this is not what we see. What we see on the screen is the pattern that would have been produced if each electron behaved at the diffraction grating not like a particle but like a wave. So we attribute a wave nature to electrons, and we say (perhaps metaphorically) that the wave packet of the electron is 'smeared out in space' when it passes the diffraction grating.

But now consider what happens when we look to see where the electron is. We do this at the recording screen, and what we see when we catch a single electron there is a single highly localized spot. As a consequence of the measurement, the electron, which was spread out in space, appears in a single place. We say that the measurement 'collapses the wave packet': the wave of the electron, which before covered a large area in space, is now peaked in a very small region.

There are two kinds of reasons for believing in the projection postulate. The first are empirical; the second, conceptual. The most primitive empirical reasons are like the one we have just been discussing: when we do carry out measurements on quantum systems, it seems we find the systems in their reduced states; but we know that they would not have been in those reduced states if we had not measured them.

A somewhat more sophisticated reason argues from our ability to prepare quantum systems in reduced states. Inagine for example that for some experiment we want to perform we need to have a beam of electrons all of which have their spins directed upwards along some axis. But electrons don't usually come with their spins lined up like that. Just as an electron my have no well-defined position, so too it may have no well-defined direction for its spin. It may be in a 'smear' of both spin-up and spin-down along the direction in question. Such a smear is called a superposition of up, down states.

What we do in a case like this is somewhat akin to measurement. We send the beam of electrons through an inhomogeneous electromagnetic field. This sorts the electrons into two channels according to whether the spin is up or down. Then an absorbing plate is put in front of the down channel. We are left with a bean which we can use in our experiment: every electron seems to have spin-up. Notice that this procedure requires a reduction of the wave packet. If we are to end up with a beam of electrons all spinning the same way, somewhere or other, either in the field or at the absorbing plate, the electrons have got to get out of their smeared state and into one of the two reduced states, spin-up or spin-down.

The second kind of reason for admitting the projecting postulate is conceptual. Quantum mechanics is a probabilistic theory. But what are the probabilities of? They cannot be probabilities for systems to have values in the ordinary way since in general quantum systems don't have values. That is the point of the diffraction example. We may calculate something which we loosely call the probability 'that the electron is in a certain small region' in the grating. But in fact we know that none of the electrons are *ever* really in any of those regions.

The projection postulate solves this problem. If a measurement of position is made (as it is at the recording screen) the state of the electron changes. At the end of the measurement the electron is located in some small region, and the probabilities in question are probabilities for that to happen. Thus, quantum mechanics gives the probabilities that a system will have particular values for observable quantities *when those quantities are measured*.

According to the projection postulate, there is a physical change in the state of a system consequent on its being measured. One of the remarkable features of this change is that it is indeterministic. Using the past history of the electron, the theory can tell only with what probability it will appear at various places on the recording screen; it cannot tell where a particular electron will actually appear. This is not surprising since it is well-known that quantum theory is an indeterministic theory. But it is important to notice that it is only via the projection postulate that indeterminism enters. The normal time evolution that quantum states undergo when a measurement is not being made is deterministic. It is governed by the Schroedinger equation, and the Schroedinger equation allows us to infer from the state at any one time plus the energy which describes the system in between times what the state at any other time will be. Without the projection postulate, nothing indeterministic ever happens in quantum mechanics.

## II. WHY WE SHOULD BELIEVE IN REDUCTION OF THE WAVE PACKET

What is wrong with the projection postulate? What is wrong is that it singles out measurements from among all other interactions and says that something different happens when a measurement occurs. Given the projection postulate, there are two very different kinds of evolution in nature: the deterministic Schroedinger evolution which occurs most of the time, and the indeterministic reduction of the wave packet which occurs just when measurements are made. But what is special about measurements? How does nature know when it is supposed to behave in one way and when in the other?

Here is where consciousness enters. No one has succeeded in giving anything like a reasonable physical characterization of which interactions are measurements and which are not. Eugene Wigner at least faces the problem squarely. For Wigner, possibly for von Neumann, and others, what is special about measurement is that ultimately it involves an interaction with consciousness, and consciousness is a very different non-physical kind of factor, not governed by the Schroedinger equation.

I think the emphasis on measurement here is mistaken, however. There are two kinds of evolution, but measurement is not the crucial distinguishing feature. Reduction of the wave packet is not confined to measurement situations. The reasons which argue for reduction of the wave packet on measurement apply equally well in other cases, most notably in situations in which damping occurs. Radioactive decay is the most familiar example. I will talk about a slightly simpler example, not the one in which a nucleus de-excites and gives off alpha or beta particles or whatever, but rather the case in which the electron of the atom changes energy levels and gives off photons. In both cases the rate of decay is exponential.

There are two kinds of stories one can tell about decay. The first, which I shall call the 'old quantum theory' story, is the one which is familiar to most of us. The probability for an atom to remain in its excited state drops off exponentially in time; eventually the atom will decay and give off a photon. The process is indeterministic: the state of the atom changes, but there is no reason why it does so at one time rather than another or why one particular atom de-excites rather than any other.

This contrasts with the new quantum theory story – the story we are led to tell by following the formal treatment of decay. Here we think of the atom-plus-electromagnetic-field as a composite system which starts out in the state *atom excited, no photons present*. Over time this state evolves in a completely deterministic way under the Schroedinger equation into a superposition of the states *atom excited, no photon present* and *atom de-excited, one photon present*. As time goes on the second state will constitute a larger and larger portion of the smear; in the limit as time goes to infinity all of the smear will be in the *de-excited plus photon present* state. But at no finite time will a single atom have ever de-excited. The two states are hopelessly intertwined. Worse, the atom and the field are also irrevocably joined. Without some reduction of the smeared state to one or the other of its constituents, it is not possible ever to represent the atom by itself again.

The two stories are quite different, but there is a trick which might reconcile them. In the end, however, the trick doesn't work unless we assume reduction of the wave packet. The trick depends on a fact about statistics for components. Since we are describing statistics, it is convenient to think in terms of large collections, say a thousand or ten thousand excited atoms. If we take the formalism literally, after a short while each atom is intertwined with its field and no longer evolves on its own. But imagine that we are interested only in the atoms in our collection and never intend to look at the photons; or alternatively, we are interested only in the photons and will never care about the atoms. In this case we can write down something we might call a 'state' for the atoms which has a very nice characteristic: for any future measurements to be made *on the atoms alone*, this state gives exactly the same statistical predictions as does the combined atom-field state which is dictated by the Schroedinger equation. There is a similar 'state' for the photons.

The difficulty with the new 'states' is that they cannot account for the behavior of the individual atoms over time, unless we assume reduction of the wave packet. The new state for the atoms gives the right statistics for the collection of atoms at any time. But it does not determine that the individual atoms behave properly over time – that the de-excited atoms stay de-excited and that the excited atoms de-excite as they should. We always get the right statistics in our collection, but we don't know that the individuals have the right kind of history.

This is exactly analogous to the problem of preparation that we discussed earlier. I didn't mention it, but we can do exactly the same trick with the spin-up spin-down electrons that we have done here. So long as we are never going to look again at the magnetic field and the absorbing screen, we can represent the electrons afterwards by a new 'state', and this new 'state' will give the right statistics for the collection of electrons at any time. It will, for example tell immediately after the absorbing plate what portion of the original beam will be *unabsorbed electrons from the upper channel with spin up*, and how many will be *absorbed electrons from the bottom channel which had spin down*. Moreover it will also tell, for any later time, that exactly the same proportion of electrons will be in the time-evolved of the *unabsorbed-upper channel-spin up* state versus the time-evolved of the *absorbed-lower channel-spin down* state. But what it will not tell is that the individual electrons contribute to the statistics in the same way. Some individual electrons which appeared as unabsorbed at $t_1$ may appear as absorbed at $t_2$, and vice versa, so long as the individuals switch in such a way as to preserve the over-all statistics. To rule this out, we have to assume that each electron behaves as if it *really is* in one of the two reduced states, but that is tantamount to assuming reduction of the wave packet for these electrons.

The same is true for our decaying atoms. If we want the individual atoms to behave correctly over time we need the story of the old quantum theory. At some time the wave-packet for the atom-field combination reduces: the atom really decays and a photon is really given off. I should make clear that this account differs empirically from the new quantum theory story; and it is not uncontroversial. I assume that the atoms genuinely decay; the new formalism suggests that they stay in a superposition forever, or at least until a measurement is made. There are in principle observations which could test the difference. But these observations require us to look at peculiar quantities on correlated atom-field pairs; and they are not ones that we know how to carry out. (Note: To test the difference between a genuine mixture of atom-field states and a superposition, we must measure some observable *on the composite*, and it must be an observable which does not commute with the questions *excited atom-photon absent*? and *de-excited atom-photon present*?)

Thus the accounts given by the old and new quantum theory are empirically different, but the difference cannot be experimentally verified in a straightforward way. I think that Hilary Putnam is right: physicists are apt to reconcile this difference in an *ad hoc* manner. They solve the problem of preparation the same way. On the one hand the formalism pictures ever larger and larger systems evolving deterministically as an inseparable whole, and the projection postulate supposes that they continue to evolve that way until a measurement is made. On the other hand, in order to account for preparations, or to tell the familiar story about exponential decay, or to deal in a sensible way with other damping phenomena, we suppose that reduction of the wave packet occurs at some specific point in the process independent of measurement. The difficulty is resolved by assuming that whatever occurs at the point when the wave packet needs to be reduced constitutes a measurement. As Putnam says about how physicists keep macroscopic observables out of superpositions:

> Most physicists are not bothered by Schroedinger's cat. [Schroedinger's cat is in a superposition of live and dead states.] They take the standpoint that the cat's being or not being electrocuted should itself be regarded as a measurement. . .. More precisely, the reduction of the wave packet takes place precisely when if it had not taken place a super position of different states of some macro-observable would have been predicted. ['A Philosopher Looks at Quantum Mechanics', p. 25 in R. Colodny, *Beyond the Edge of Certainty*, Pittsburgh Studies, Vol. I.]

Measurement here is obviously not being used in any serious sense. If I am correct about what happens in exponential decay, or in cases of state preparation, we have plenty of good examples of reduction of the wave packet without the intervention of an observer. The decaying atom changes its states spontaneously and indeterministically, and no measurement is required to precipitate the change. There is no need for consciousness in quantum mechanics.

Nevertheless, the basic problem with the projection postulate remains. There are two kinds of evolution in quantum mechanics: the deterministic evolution dictated by the Schroedinger equation, and spontaneous reduction of the wave packet in which the final states are not determined by the previous history but only their probabilities are determined. How does nature know which pattern it is to follow?

## III. WHY WE SHOULD NOT BELIEVE IN THE MEASUREMENT PROBLEM

This was the state of my thinking a couple of years ago. In certain circumstances – radioactive decay is an example – quantum systems make transitions from one state to another and without the intervention of measurement. This accounts for the possibility of preparation, and it allows us to tell sensible stories about damping phenomena. Moreover, it provides a coherent way to interpret quantum theory. I asked earlier, "What are the probabilities in quantum mechanics probabilities of?" So long as we admit only transition probabilities, the answer is straightforward. They are probabilities for a system to make a given change of state in one of these special situations. These probabilities are perfectly classical: the events over which they are defined have a classical logic; and no consciousness enters.

I became discouraged, however, because it seemed impossible to provide a realistic characterization of the circumstances in which Schroedinger evolution failed and reduction of the wave packet occurred. The situations in which it seems most certain that reduction occurs are all irreversible situations, modelled in the theory by an interaction with a system which has an infinite number of degrees of freedom. There is good reason for this. If things are sufficiently random, when we average over the infinite degrees of freedom their effects will cancel each other, and the interference terms which are the peculiar marks of a superposition will disappear.

But this distinction does not provide a realistic difference for nature to operate on. In reality, all systems have a finite number of degrees of freedom.

Schroedinger evolution and reduction of the wave packet are very different kinds of evolution. Sheer size cannot be a good mark of the difference. It would be a foolish question to ask, "Exactly when is a system big enough for nature to think it is infinite?"

I now think that this whole characterization problem, which has been so vexing, is a psuedoproblem. It is an artifact of the mathematics. There are, however, related problems which are genuine, and are quite hard to solve. Most broadly described, these are problems of how to decide which is the right theoretical description for a given physical situation. But these are problems for which there is no general solution, just the piece by piece work of everyday physics. The measurement problem in quantum mechanics is usually seen as a puzzling conceptual problem, a problem which exposes an important incoherence in the theory but which nevertheless does not seem to stand in the way of the highly detailed and successful development and application of the theory. It now seems to me, to the contrary, that there is no conceptual problem, there are just hard physics problems.

There is no characterization problem because there are not two different kinds of evolution in quantum mechanics. There are evolutions which are correctly described by the Schroedinger equation, and there are evolutions which are correctly described by something like the projection postulate. But these are not different kinds in any physically relevant sense. We think they are because of the way we write the theory.

I have come to see this by looking at theoretical frameworks recently developed in quantum statistical mechanics; E.B. Davies book, *Quantum Theory of Open Systems*, probably represents the best abstract formalization. (Note: I should note that the use which I want to make of Davies' formalism is probably not one of which he would approve. Moreover the fact about the formalism which I stress is an insignificant part of this very interesting book.) But the point is simple to see and does not depend on any of the details of the statistical theory. Von Neumann said there were two kinds of evolution; and he wrote down two very different looking equations. But the framework he provides is not a convenient one for studying dissipative systems, which are the primary focus of statistical mechanics. As I discussed earlier, in the cases of damping we see behavior which looks like reduction of the wave packet. The Schroedinger equation is not well suited to handle these. On the other hand, the projection postulate is not rich enough to allow us to give detailed treatments of physical phenomena involving damping, such as lasers. So quantum statistical mechanics has developed a more abstract formalism which is better suited to the problems it treats. This formalism writes down one kind of evolution equation, of which the Schroedinger equation and the projection postulate are special cases.

The evolution prescribed in the quantum statistical formalism is very much like Schroedinger evolution but with one central difference: the evolution operator lacks a certain mathematical characteristic, the characteristic of unitarity. An evolution operator is, essentially, an operator (positive, linear, trace-preserving) indexed by time, which takes states (density matrices) at one time into states at some other time. To be unitary means that it preserves the lengths of vectors and the angles between them. So the new formalism gives up unitarity. What is the physical significance of unitarity? The primary effect of keeping vector lengths and angles fixed is to stop reduction of the wave packet. In quantum statistical mechanics, the evolution operators retain their other important characteristics, but they need not be unitary, and so reduction of the wave packet is permitted.

In this new formalism there is one equation, but a distinction can still be drawn. All situations to which the Schroedinger equation applies are situations in which the evolution operator is unitary; cases to which the projection postulate applies are among those in which the evolution operator is non-unitary. But this does not mean there are two kinds of evolution. Unitarity is an interesting mathematical characteristic of operators: an operator is unitary if its adjoint is its inverse. But this does not mean that there is any real physical characteristic which obtains in all and only those situations to which unitary operators apply. We have been misled by taking the mathematics too seriously in this case. Obviously not every interesting mathematical distinction in a representation is going to mark a physical distinction in the things represented. This is the case with unitarity.

Looked at from the point of view of the quantum statistical formalism, there is only one equation for nature to obey. We still might ask, "For a given situation, how does nature know whether to evolve the system with a unitary operator or with a non-unitary operator?" But this is the wrong question. The appropriate question is "How does nature know how to evolve the system?" The answer to this question is that she looks at the forces, then does what is appropriate. Sometimes the forces are represented in our theory by operators which happen to be unitary; sometimes not. But this makes no difference in nature.

I sound as if unitarity has no significance other than mathematical interest. Is this so? No, because unitarity is just the characteristic that precludes reduction of the wave packet; and, as we saw, reduction of the wave packet is indeterministic, whereas Schroedinger evolution is deterministic. What I want to claim is that there need be no further general physical characteristic which is true of those situations in which the evolution is deterministic, and which fails to be true when the evolution is indeterministic. The forces which obtain in a given situation determine how the system will evolve, given the quantum statistical equation. Once the forces are fixed, that fixes whether the evolution of the system is deterministic or indeterministic. We don't need any further physical facts about the forces to fix this.

Unitarity plays a purely theoretical role. We are interested in determinism versus indeterminism, and this should be reflected in our theoretical representations. We look for some characteristic of the representation to mark the distinction between evolutions which are deterministic and those which are indeterministic. Unitarity does this job. But we must not take it to represent some further physical factor, a factor in virtue of which all deterministic situations will be deterministic and all others will be indeterministic. Unitarity marks a distinction we are interested in. It does not pick out a physical basis for that distinction.

We can see now why I said that related problems remain, but that they require solutions in physics, not philosophy. We must still figure out what forces apply in any given situation. But this is what ongoing physics is all about; and there is no general procedure for how to proceed. We use our physical intuitions, analogies with other cases, specializations of more general considerations and so forth; sometimes even we choose the models we do because the functions we write down are ones we can solve. As Merzbacher remarks about the Schroedinger equation:

> Quantum dynamics contains no general prescription for the construction of the operator $H$ whose existence it asserts. The Hamiltonian operator must be found on the basis of experience, using the clues provided by the classical description, if one is available. Physical insight is required to make a judicious choice of the operators to be used in the description of the system (such as coordinates, momenta, spin variables, etc.) and to construct the Hamiltonian in terms of these variables. (Eugen Merzbacher, *Quantum Mechanics*, 2nd ed., pp. 336–337.)

As Merzbacher says, physical insight is required to choose the right operators; but this choice is all there is to the measurement problem.

## NOTE

* This paper will appear in an expanded version in a collection of the author's papers on Realism: Nancy Cartwright, *How the Laws of Physics Lie*, Oxford University Press (expected publication date: Jan. 1982).

JEREMY BUTTERFIELD

# MEASUREMENT, UNITARITY, AND LAWS

In the first section of this paper, I first describe the measurement problem as a background to discussing Cartwright's approach to it. This discussion prompts an examination of her general views on theoretical laws, theoretical entities and explanation, as expounded in her recent papers (1980, 1980a, 1980b, 1980c). So in the second section, I report and then discuss these views.

## 1. QUANTUM MECHANICS

### 1. *The Measurement Problem*

I think about this problem in much the same way as Cartwright. But it will forward the discussion, I think, to give my perspective on it.

Quantum mechanics associates states with physical systems. We can take these states to be given by vectors in a vector space. They are associated with the results of measuring a quantity – say energy – in a rather indirect way. Each quantity, $H$ say, is associated with a function or operator, mapping vectors to other vectors. This operator, also called $H$, has eigenvectors, i.e., vectors whose image under $H$ is a multiple of themselves, $Hv = \lambda v$ ($\lambda$ a number). The eigenvalues, $\lambda$, are the possible results of measuring the quantity. The probability of getting a result $\lambda$ on measuring a quantity on a system with state-vector $w$, say, is proportional to the 'closeness' of $w$ to the eigenvector associated with $\lambda$. More precisely: if $\{v_i\}$ are the eigenvectors associated with a quantity and $\{\lambda_i\}$ are their eigenvalues, and if the state of the system $w$ is expressed in terms of these eigenvectors as $\sum_i c_i v_i$, $c_i$ some constants, then: the probability of getting result $\lambda_j$ on measurement is $|c_j|^2/\sum_i |c_i|^2$. It follows that if the state is an eigenvector, $w = v_j$, measurement will give the result $\lambda_j$ with probability 1.

This talk of probabilities suggests that to ascribe that state $w = \sum_i c_i v_i$ is to say that the system has some definite value $\lambda_j$ for the quantity associated with the operator $H$ – but in general we can only assign probabilities to various values. And this view – that probabilities attach to our ignorance – seems supported by the fact that for some types of measurement, an



*R. Swinburne (ed.), Space, Time and Causality*, 135–147.

immediate repetition of the measurement gives whatever result was first observed. Here it seems the first measurement relieves our ignorance, but does not interfere with the value. Let us call such measurements ideal.

However, there are great difficulties with this view. The 'no hidden variables' proofs give strong evidence against it. But objections can obviously be raised at a humbler level, say by considering the uncertainty relations. If systems possess definite values for quantities, why can't we prepare collections of them with definite position and momentum?

Accordingly the traditional view is that quantum systems do not generally have a definite value for each quantity; and quantum mechanical probabilities attach to measurement results, not possessed values. This view agrees however that if the state is an eigenvector of the measured quantity, so that the corresponding eigenvalue has probability 1, then that value is possessed. What of ideal measurements? Here we want to ascribe to systems, as their state after the first measurement, an eigenvector of the measured quantity, viz. one whose eigenvalue is the result of that measurement. So systems that had the same pre-measurement state are now given different states. The entire collection of systems can be ascribed, as a post-measurement state, a so-called mixture or mixed state; this is essentially a weighted sum of the various eigenvectors, the weights being the probabilities of the various results in the first measurement. But this sum is not like adding vectors in the vector space we started with. This mixed state is quite different from the initial state-vector $w = \sum_i c_i v_i$ – also called a pure state. Though the two states agree on the probabilities for results of measuring the quantity in question, there are other quantities for which they give different probabilities.

The crucial point about ideal measurements, and the mixed states we ascribe after them, is that: the evolution from the initial pure state to the mixed state cannot be obtained from the Schrödinger equation. This describes a so-called unitary evolution of isolated systems, which sends pure states (i.e., state-vectors) into other pure states. To send a pure state to a mixed state, we need another type of evolution. The projection postulate provides it, by postulation: in ideal measurements, pure states go to the corresponding mixtures. This is the classical form of 'reduction of the wave packet'.

Accounting for the results of repeated ideal measurements is not the only application of the projection postulate. It promises to reconcile quantum systems' apparent lack of definite values for quantities with macroscopic objects' apparent possession of them. The idea will be this: even when we make the value of a macro-quantity depend on a micro-quantity which may have no definite value, somewhere in the relevant clain of events the projection postulate will send a pure state to a mixture, and prevent indefiniteness of a macro-quantity.

The problem is that we lack a principle about when the discontinuous evolution of the projection postulate, as opposed to Schrödinger evolution, occurs. As Cartwright says, some have appealed to consciousness for a discontinuous evolution to a mixture. But if we reject this, one obvious suggestion is that there is in fact no such evolution, i.e., that it is an artifact due to considering as isolated a system which is, or has been, in interaction with another. After all, Schrödinger evolution is supposed to govern isolated systems; and in measurement the quantum system is not isolated. This suggestion can apparently be supported by the quantum mechanical formalism for discussing interacting systems. Given a pure state of a composite system, we can write down a 'state' for each of the component interacting systems, which, as Cartwright says, gives the same predictions for any measurement on the component system alone as does the pure state of the composite. This 'state' is mathematically just like a mixture; in fact it is usually called a mixture. But unfortunately, it cannot be interpreted as above; that is, we cannot say that each component system is in one of the pure states that figure in the weighted sum given by this 'state'. For if we said that, the composite system would be in a mixed, not pure, state.

One might hope that despite this *general* difficulty, one can secure a definite value for macro-quantities of systems interacting with microsystems. That is, one might hope that with a suitable notion of measurement-interaction, Schrödinger evolution of a composite system comprising a measuring apparatus and a measured system will secure a definite value for the apparatus quantity, 'pointer position', that is supposed to depend on a micro-quantity of the measured system. But there are powerful negative results here. They run along the following lines: one gives a definition of a measurement-interaction between an apparatus and a measured system; this definition treats the systems quantum mechanically, and must allow information about the micro-quantity of the measured system to be inferred after the interaction from an apparatus quantity, 'pointer position' $P$. Then one shows that there are no interactions of this sort that secure a definite value for $P$. (Cf., for example, Fine, 1970.)

These negative results renew one's interest in hidden variables. After all, if we can't get the mixed states we desire by a *principled* use of the projection postulate, or by 'mentally separating' one of two interacting systems, perhaps we had better ascribe values to systems when their states are not eigenvectors of the corresponding quantity. Thus the hope will be that we never need the projection postulate: all evolution is Schrödinger evolution but that does not prevent quantities having definite values. This approach involves getting around the 'no hidden variables' proofs, and around Bell's theorem. It also involves arguing that the pure states one is now happy to ascribe to macro-systems predict the right statistics for results of measurement on them. Cartwright (1974) noticed that in this argument, one could use the work of Daneri etc. on the indistinguishability, as regards statistics for a broad range of measurements, of pure states and mixtures. (As Cartwright says, 'subtle' experiments we can't now perform could distinguish them.)

## 2. *Cartwright's Approach and Comments*

Cartwright does not endorse this sort of hidden variables approach. She believes we need the projection postulate; and certainly it is difficult to get around the 'no hidden variables' proofs and Bell's theorem. She also presents a difficulty about individual histories. Though the observed statistics of a broad range of measurements may be recovered by the pure state as well as by the mixture prescribed by the projection postulate, only the latter can explain how each individual system contributes to the statistics. Thus only the latter can explain why on repeating an ideal measurement a system gives whatever result it gave before. This is certainly a difficulty. It is not of course a knock-down argument; hidden variable theorists can hope to account for individual histories in terms of the assignments of values to quantities with which they supplement the pure states.

But doesn't accepting the projection postulate mean Cartwright has to give some principle about when it applies? She thinks not, because we can see both Schrödinger (i.e., unitary) evolution and evolution by the projection postulate as special cases of a more general kind of evolution. Quantum statistical mechanics has provided a general theory of evolution of quantum states (pure and mixed) that encompasses these two sorts of evolution, and many others, as special cases. Nor is this just a mathematical umbrella. It allows us to set up detailed models of phenomena that cannot be treated easily, if at all, with the traditional formalism's two sorts of evolution.

There is no problem of characterizing when these different sorts of evolution occur, because there is only one *kind* of evolution in quantum mechanics, viz. the most general sort of evolution; the detailed physics of particular cases will prescribe what form this takes. This means in particular that unitarity does not pick out a physical property of the evolution. Nor does determinism: this is a more general feature than unitarity – I presume it includes antiunitary evolution. But it does not exhaust the general sort of evolution; this includes non-deterministic evolutions like the projection postulate's. Of course we may be interested in determinism or unitarity, and single out these features in expounding the theory; but that doesn't mean some physical property is true of exactly the situations where the evolution is deterministic or unitary.

I think Cartwright's approach is exciting. My reasons are clear enough. I would rather not appeal to consciousness to provide a mixture. And while I want macro-quantities to have definite values, I am sceptical enough about the prospects of evading the 'no hidden variables' proofs and Bell's theorem to want a mixture. The traditional difficulty with this line is the characterization problem. So it's marvellous to be told this is a pseudo-problem. If we have to say that no physical characteristic co-varies with such general features of the evolution as determinism or unitarity, so be it. Even if we previously wanted to deny this, it seems a small price to pay for the liberation.

But I doubt that we have to say this. That is, I see no guarantee that features like unitarity and determinism are present in so large and heterogeneous a body of circumstances that we cannot speak of a corresponding physical property. And in order to deny that there is a *conceptual* problem of measurement, Cartwright does not need such a guarantee. She only needs that if physical properties correspond to these features, they can only be revealed by examining the physical details of interactions. Actually, I think Cartwright would agree with this. She makes precise the idea of there being no corresponding property by saying that the forces, i.e., the physical details, determine whether the evolution is deterministic. And that is compatible with determinism's being present in a sufficiently homogeneous body of circumstances for us to talk of a property. Indeed, on some less stringent existence-criteria for properties, such determination would *imply* that determinism is a property.

However, in allowing that a hypothetically isolated system may evolve in a nonunitary way, Cartwright's view is more radical than might at first appear. For Wigner's theorem gives unitarity a more central role than her remarks suggest. Roughly speaking, this theorem states: suppose a system's time-evolution is independent of the time-origin and is given by a continuous homomorphism of the reals into the automorphisms of the system's lattice of propositions; then its evolution is induced by a one-parameter group of unitary maps on the Hilbert space (cf. Jauch, 1968: Sections 8-2, 8-3, 9-4, 9-5, 10-2). I suppose the assumption that Cartwright would be most inclined to reject is that evolution of an isolated system should be given by an automorphism of its propositional lattice; and this will no doubt correspond to some sort of loss of information in the system. I see no knockdown objection to such a rejection, but I want to stress that it is quite radical.

Yet it is natural to ask whether Cartwright needs to allow nonunitary evolution for isolated systems. Might we not use the more general formalism provided by quantum statistical mechanics to reconcile unitary evolution for isolated systems with Cartwright's advocacy of nonunitary evolutions for measurement interactions? Davies (1976) is in fact concerned to reconcile the general evolution he discusses with a unitary evolution of a larger system of which the system being modelled is a component. Thus he gives existence-proofs, for certain kinds of nonunitary evolution, of a larger space of state-vectors, on which some unitary evolution produces the given (nonunitary) evolution (1976, Chapter 9). He also considers deriving nonunitary evolutions of component systems from a given unitary evolution on a composite system (1976, Chapter 10). Might not Davies' more general formalism avoid the objection we raised on p. 3 against regarding the projection postulate as an artifact of considering a non-isolated system? Unfortunately, the formalism is not general in the right kind of way.

As we are considering unitary joint evolution, Davies' general notions of operation and instrument are not immediately relevant. His generalized notion of an observable as a positive operator-valued measure, rather than as a projector-valued measure, does not prevent our being led by the usual considerations to the partial tracing condition for component states; and if all projectors are observable, component states will be determined in the usual way, so that p. 137's problem of 'improper mixtures' will arise. Similarly the impossibility proofs mentioned on p. 137 will not be blocked. From this perspective, it seems that if we do not want to supplement quantum states with extra value-ascriptions, we should follow Cartwright in allowing nonunitary isolated evolution.

My major comment on Cartwright's approach arises from asking more exactly what this more general formalism provides. (I will follow Davies, 1976.) In this formalism one can describe models in which pure states evolve to mixtures; and apply these to actual phenomena. But I see no direct evidence that these can be applied to situations we call measurements; or that the mixtures obtained can be interpreted in the right way, viz. as representing a collection of systems each member of which has a definite value for the relevant quantity. (In fact I see no evidence that the mixture is of the right sort mathematically for this interpretation to be possible.)

I do *not* want to pour cold water on this programme; I find it very attractive. But I want to stress that it *is* a programme, not a *fait accompli*. To succeed with it, we need to provide detailed analyses of measurement situations, showing that the right mixtures are forthcoming. We need not of course cover all measurement situations, but we need to make it plausible that the right mixtures are generally obtained. Here 'generally' need not mean 'universally'; it is the pervasiveness, not necessarily universality, of definite values that needs to be explained. Only when we have such detailed analyses will Schrödinger's cat be laid to rest. (And as philosophers we should remember that even if there is thus no philosophical or conceptual problem of measurement, there are still philosophical issues, e.g., of ontology, connected with non-locality.)

However, even if I were capable, this would not be the place to contribute to the details of this programme. I want instead to examine a philosophical motivation that contributes to Cartwright's attraction to this programme. I mean her distrust of theoretical laws, as expounded in her (1980, 1980a, 1980b, 1980c). So in the next section, I report and then discuss this motivation. There is at least this connection between her advocacy of the above programme and her general views: someone who distrusts theoretical laws will generally be happier than the rest of us with logically weak covering laws, like a law for a system's evolution that leaves much to be determined by circumstances. And I will hold there may be other connections.

## 2. THEORETICAL LAWS, THEORETICAL ENTITIES, AND EXPLANATIONS

### 1. *Cartwright's Views*

Cartwright is suspicious of theoretical laws. She agrees with van Fraassen (1980) in finding inference to the best explanation a suspect procedure: no matter how well – or how much better than its rivals – a hypothesis explains or organizes or unifies a broad range of phenomena, that in itself gives us no reason to believe it.

She also believes that many or most of our theoretical laws are false – and now known to be false. However, that need not prevent their being explanatory: for there is often a trade-off between truth and explanatory power. The picture is this (Cartwright, 1980, 1980a). The obvious formulations of such laws as Newton's law of universal gravitation or Coulomb's law of electrostatic attraction, are false. To be true they require some sort of *ceteris paribus* clause; in these examples, the obvious rider is that no other forces are present. But such a clause is rarely satisfied, so that the amended law applies to very little. So far we can still maintain that our theoretical laws are true; we just have to be modest about their scope. But this is where the trade-off between truth and explanation comes in. We often explain phenomena by 'composition of causes', i.e., by taking several laws to act in combination; and this requires that the laws describe the same effects whether or not other laws apply. But if they do this, they will truly describe the phenomena only in rare cases where other laws which would modify the effect do not apply. Thus the explanatory power of the law of universal gravitation depends on the fact that the gravitational force between two objects is independent of whether or not they are charged (outside general relativity anyway). But only if they are uncharged can it describe the actual phenomena.

An obvious reply to this is that the effects described by individual laws *are* present in the 'combined case'; in the example, the gravitational and electrostatic forces just add vectorially to give a total force. Cartwright doesn't think this reply works. I think it does, in this example; but I shall not argue this here. For Cartwright goes on to provide a more plausible example where such a decomposition of the total effect is allegedly impossible. The best one can do in this case, by way of a true claim about the operation of the Coulomb force, is supposed to be a counterfactual 'If it alone were at work, then . . . '. But, says Cartwright, we have no model of explanation showing how this counterfactual bears on the explanation of the phenomenon in question (viz. quantum mechanical level-splitting); in particular, the covering law model allows the relevance of actual, but not of possible, facts.

Cartwright of course admits that composition of causes is not the only sort of explanation; and other sorts, such as covering law explanation, are obviously compatible with laws' truly describing actual phenomena (and in some versions demand it). Furthermore, there is sometimes a single covering law for cases of composition of causes, e.g., the case of gravitation and electrodynamics. But Cartwright stresses that such covering laws are very often not available; and we should be able to make sense – as explanations – of the explanations we provide in their absence. Even when they are available, Cartwright thinks they do not tell 'the whole story'; for only by looking at the combined action of the separate laws can we see the detailed causal story leading to the phenomena. So, says Cartwright, the known falsity, or very small scope, of these laws would still have to be faced.

Composition of causes is not the only case where covering law explanations are scarce. There is also the vast number of cases where we explain by means of a single law whose *ceteris paribus* clause is known not to be satisfied. For example, Snell's law is valid only for optically isotropic media. But we use it to explain that the refraction angles in a mildly anisotropic medium are close to those given by the law. Cartwright stresses that this is a *decision*, albeit one for which there might be good reasons, e.g., some general – and unformulated! – assumption about continuity.

The covering law theorist will say that in these latter cases, we are not explaining, but betting on the form taken by the true law – and so by the correct explanation. But Cartwright objects that this assumes there are laws, albeit usually unknown ones, that cover every phenomenon; and explanation ought to make sense without that assumption. It also implies that most of the time when we think we're explaining, we're not – we're betting on the existence and form of an explanation; and this seems wrong.

So much by way of summarizing Cartwright's view of theoretical laws and explanation. However, she does believe in theoretical, i.e., unobservable, entities. In (1980b) she holds that as these figure in our causal explanations of phenomena, we are justified in believing in them. For believing a causal explanation involves believing in the cited cause; so in the causal case, 'explanation leads to inference' – but an inference to the existence of an object, not a law. As an example, she cites inferring the unobservable electron as the cause of the tracks in the cloud chamber. (Cf. also 1980c.)

## 2. *Comments*

I shall begin with theoretical entities. One obvious motivation for accepting these is that once one admits there is a spectrum of observability, it is surely difficult to deny that we could be justified in believing in entities of even the most unobservable sort. If one accepts this, but is attracted by other aspects of 'anti-realism' – say, by scepticism about inference to the best explanation – Cartwright's endorsement of theoretical entities but not theoretical laws is bound to seem attractive.

Of course some theoretical entities might not be redeemable by their being part of the causal network; a possible example is spacetime points or the metric field in a theory in which spacetime is not 'dynamical'. But one may well be happy not to believe in such causally inert entities. As I see it, the problem with Cartwright's suggestion is that the theoretical entities we do want to believe in may not figure in acceptable causal explanations, in a way that justifies our belief. Thus if causation is a relation between events – as it is normally taken to be – I can infer from a causal explanation the existence of an event, viz. the cause. But there seems no guarantee that belief in the cause will justify belief in the theoretical entity of interest. At least this is so, unless we consider causal explanations that involve an appropriately large part of theory. For the beliefs Cartwright admits, viz. beliefs about theoretical entities' causal relations to the phenomena, are not rich enough to single out our theoretical entities. Or if they are, they bring with them beliefs about causal relations among such entities, or more generally theoretical laws.

Thus in the cloud chamber example, the cited cause may be 'condensation', 'condensation due to ionization', 'ionization due to a passing electron', or a detailed description of a typical electron-atom interaction. Each of these descriptions could figure in a causal explanation of the tracks. But could we believe the explanation that describes the cause in terms of electrons, without also believing a substantial body of theory about electrons? I doubt it: Cartwright needs an argument that causal explanation only commits one to suitably 'low-level' laws.

I turn to Cartwright's views on laws and explanation. There is a lot here with which I agree. It is salutary to emphasise that covering laws are generally scarce and, once one adds an appropriate *ceteris paribus* clause, of small scope. And it is important to note that when we apply laws without these clauses being satisfied, we are taking a decision, not forced on us by what we know. But what about the claims, made in this connection, that the covering law theorist is liable (a) to assume that there is a law (usually unknown!) covering every phenomenon; and (b) to construe much of what we do by way of explanation as explanation-sketches or promissory notes?

I think the covering law theorist will point out that there is an obvious danger of a verbal dispute here: his explanation-sketch or promissory note may well be Cartwright's explanation. And he need only make assumption (a) if he thinks every phenomenon can be explained – a strong claim, in his sense of explanation. Admittedly, he must impute to someone who uses Snell's law to explain the approximate refraction angles in a mildly anisotropic medium, a belief that there is an unknown law covering the phenomenon; but this seems reasonable. Of course, we need an account – presumably using the idea of approximate truth – of what the covering law theorist calls explanation-sketches or promissory notes. But so far I see no reason to think he is ill-equipped to start giving such an account.

Admittedly, given that covering laws are generally scarce and of small scope, inference to the best explanation will rarely be possible. We will rarely have an explanation we consider best – whatever we exactly mean by that. But the covering law theorist, and more generally the scientific realist, obviously do well to accept this. It should be no part of either doctrine that we believe some explanation of every phenomenon; that is quite different from believing those (best) explanations we do have. This remains true even if we add to scientific realism in van Fraassen's (1980) sense – that scientific theories aim to be literally true accounts of the world, and that to accept them is to believe them – the assumptions that we always accept some theory and that explanation is closely tied to a theory's descriptive resources. For we are not provided by our belief in some theory with even a description of everything, except in the minimal sense of a classification in terms of our theory's predicates. Thus the realist in van Fraassen's sense can accept the scarcity of laws and explanations, while maintaining what I take to be his strongest ground: the difficulty of making sense of scientists' attitude to an accepted theory, except as belief; and the anti-realist's reliance on a suspect theoretical/observable distinction.

A similar point emerges from van Fraassen's treatment of explanation. Van Fraassen reviews the difficulties facing traditional attempts (e.g., Hempel and Salmon) to analyse explanation in terms of the straight-forwardly descriptive resources of a theory. He suggests (1980, p. 154) that because of these difficulties, explanatory power came to be seen as a special and irreducible feature of theories: and as science sought understanding, and thus explanation, one demanded more of a theory than that it 'save the phenomena'. At the same time, explanatory power was seen as a special sort of evidence for a theory. Van Fraassen offers us instead a compelling treatment of explanation as a context-dependent activity, using only a theory's descriptive resources: the context-dependence allows him to avoid the traditional difficulties. I think the correct realist response is to reject the line of thought van Fraassen suggests as characteristic of recent realist thinking; indeed, I suspect it is a caricature. The realist can also accept van Fraassen's attractive approach to explanation. All he loses as an argument for his position is a vague appeal to explanation or understanding – an appeal that he should not have respected, if indeed he did. His strongest ground remains intact.

The remaining issue is whether Cartwright's views about explanation by composition of causes are right. There are two questions: is there really a trade-off between explanation and truth in composition of causes? And, when there is a covering law, do we need the combination of separate laws to catch the 'causal story'? I am agnostic about the second question, partly because 'causal story' needs clarification. Apart from that I suspect we can recover the causal story from covering laws that combine different physical interactions, like gravitation and electromagnetism; but these may be special cases.

I think Cartwright is essentially right about the first question, and thus the realist even in van Fraassen's weak sense is challenged by this topic – but I'm not convinced he's defeated. For one thing, there is the issue whether there is a genuine case of composition of causes; I mentioned my doubts about the vector addition case. But in cases where we have no covering law, and the best we can do by way of stating how the separate laws apply to the case is by a counterfactual (cf. p. 142 above), we shall certainly have explanation in Cartwright's relatively wide sense without any laws 'stating the facts'. It is to Cartwright's great credit that she has pinpointed here a form of explanation (in a wide sense) which offers the realist a specific challenge. But even if there are such cases, the realist has some hope. He may distinguish two sorts of application for a law: in such cases we believe the law applies in some sense that justifies our acceptance of the counterfactual, but not in the sense that the effect it describes is present. This is very vague. But if it works, accepting an explanation, even one by composition of causes, will mean believing in the laws used – and that will have the advantage of promising to explain the acceptance of the counterfactual. If this line doesn't work, the realist may hope that genuine cases of composition of causes are sufficiently few that his claim that to accept a theory is to believe it is widely, if not universally, tenable. And he and the anti-realist have the joint task of determining what in such cases the attitude to laws might be.

At the end of Section 1, I noted a possible connection between Cartwright's distrust of theoretical laws and her approach to the measurement problem. It is plainly a loose one: a logically weak evolution law is still apparently a covering law. Someone inclined to scientific realism could find Cartwright's measurement programme attractive.

But looking at her views about laws in detail, we can see there may be other connections. She hints that some phenomena may be governed by no law. And if the realist response I have just suggested (and any other) cannot be made out, genuine cases of composition of causes will give examples of this in one precise sense. If quantum mechanical interactions provide such examples, the claim that such features as unitarity and determinism correspond to no property will take on a new character. It won't just be a question of whether such features occur so heterogeneously that we cannot speak of a corresponding property. There will also be examples where the evolution has a certain character – such a feature is present (or absent) –

but this cannot readily be regarded as due to the satisfaction of some jointly sufficient conditions for its presence (or absence). For these conditions' individual effects are absent. But this is *not*, of course, a reason for thinking it unnecessary to provide detailed models of measurement: that is still incumbent on us.*

## NOTE

* Many thanks to Nancy Cartwright, Mary Hesse, and David Malament for comments on an earlier version.

## REFERENCES

Cartwright, N.: 1974, 'Superposition and Macroscopic Observation', *Synthese* **29**, 229 *et seq.*

Cartwright, N.: 1980, 'Do the Laws of Physics State the Facts?', *Pacific Phil. l. Quarterly* **61**, 75–84.

Cartwright, N.: 1980a, 'The Truth Doesn't Explain Much', *Amer Phil. l. Quarterly* **17**, 159–163.

Cartwright, N.: 1980b, 'When Explanation Leads to Inference', unpublished; presented at West Washington Philosophy Conference.

Cartwright, N.: 1980c, 'The Reality of Causes in a World of Instrumental Laws', *PSA 1980*, Volume 2, forthcoming.

Davies, E. B.: 1976, *Quantum Theory of Open Systems*, Academic, New York.

Fine, A.: 1970, 'Insolubility of the Quantum Measurement Problem', *Physical Review* **2D**, 2783–2787.

Jauch, J. M.: 1968, *Foundations of Quantum Mechanics*, Addison-Wesley, Reading, Mass.

van Fraassen, B.: 1980, *The Scientific Image*, Clarendon, Oxford.

# CAUSALITY, RELATIVITY, AND THE EINSTEIN–PODOLSKY–ROSEN PARADOX

MICHAEL REDHEAD

# NONLOCALITY AND PEACEFUL COEXISTENCE

## 1. INTRODUCTION

The object of the present paper is to investigate the putative tension between the requirements of causality and special relativity (SR) on the one hand and the quantum-mechanical (QM) description of the Einstein–Podolsky–Rosen (EPR) thought experiment on the other. A preliminary formulation of the situation is that the EPR phenomenon involves an instantaneous propagation of a physical effect between two spatially separated systems which contradicts the requirements of SR, and moreover would lead to backward causation in appropriately moving reference frames. We will try to unravel this state of affairs in two stages. First we shall ask in what sense, if any, the transmission of a physical effect is involved in the EPR type of experiment and secondly whether SR definitively rules out such a possibility. This second step in the analysis will involve looking at the relationship between SR and the causal theory of time which is seen by writers such as Reichenbach and Grünbaum as providing the ontological underpinning for the clock-synchronization procedures introduced in SR in its usual standard formulation.

As we shall see the problems posed by EPR are rather less simple than the initial naive statement given above might suggest.

## 2. THE EPR ARGUMENT

In their original paper Einstein *et al.* (1935) offered an argument for the incompleteness of quantum mechanics. The example used by them involved position and momentum measurements on two spatially separated particles in an appropriately chosen quantum state. The example was not entirely satisfactory since the particular properties of the wave function used in the argument were not invariant under time-development of the two-particle system [1]. An improved and indeed simpler example of the EPR argument was given by Bohm (1951). The idea is as follows. Two spin-$\frac{1}{2}$ particles separate in a singlet state of the total spin wave function. Denoting the spin eigenfunctions for the $Z$-component $S_z$ of the spin of an individual particle with eigenvalues $+\frac{1}{2}\hbar$ and $-\frac{1}{2}\hbar$ by $\alpha$ and $\beta$ respectively, the wave function for the combined system is



*R. Swinburne (ed.), Space, Time and Causality*, 151–189.

$$(1) \qquad \Psi_{\text{singlet}} = \frac{1}{\sqrt{2}} (\alpha(1) \cdot \beta(2) - \beta(1) \cdot \alpha(2)),$$

where we label the particles 1 and 2.[2] we note first of all that the state $\Psi_{\text{singlet}}$, being spherically symmetric, can equally well be written in the form

$$(2) \qquad \Psi_{\text{singlet}} = \frac{1}{\sqrt{2}} (\gamma(1) \cdot \delta(2) - \delta(1) \cdot \gamma(2)),$$

where $\gamma$ and $\delta$ are spin wave functions for eigenstates of the spin component $S_X$ of an individual particle with eigenvalues $+\frac{1}{2}\hbar$ and $-\frac{1}{2}\hbar$ parallel to the $X$-axis of our coordinate system.

Now suppose at time $t_1$ we perform a measurement of the $Z$-component of spin on particle 1, i.e., we measure $S_Z(1)$. Suppose the result is $+\frac{1}{2}\hbar$. We infer from the mirror-image correlations built into the state (1) that if we now measure $S_Z(2)$ at a later time $t_2$ we shall find the result $-\frac{1}{2}\hbar$. Similarly if the result of measuring $S_Z(1)$ is $-\frac{1}{2}\hbar$ we can predict the restult of measuring $S_Z(2)$ at $t_2$ as $+\frac{1}{2}\hbar$. If we now apply the famous EPR *sufficient* condition for an element of reality, then if at $t_1$ we predicted the result of the measurement of $S_Z(2)$ at $t_2$, so at $t_1$ there must already exist an element of reality corresponding to the $Z$-component of spin for particle 2, and furthermore if there was no physical disturbance of particle 2 involved in the measurement at $t_1$ on particle 1, then this element of reality must also have existed at a time, say $t_3$, prior to $t_1$.

If the argument is now repeated for the $X$-component of spin using the alternative expansion (2) for the state $\Psi_{\text{singlet}}$ we conclude that at $t_3$, i.e., before measurements have been carried out on *either* particle, there must exist an element of reality associated with the $X$-component of the spin of particle 2. But orthodox quantum mechanics makes no reference to simultaneous sharp values for the noncommuting observables $S_Z(2)$ and $S_X(2)$. Hence, applying the *necessary* condition for completeness formulated by EPR, that completeness of a theory implies associating some element of the theory with every element of reality, we conclude that QM is an incomplete theory.

Before explaining the relevance of this argument to questions about causality I may remark that the argument for incompleteness does *not* really depend on considering alternative possibilities of measuring $S_Z(1)$ *or* $S_X(1)$ on particle 1. The fact that by measuring $S_Z(1)$ alone we can show the existence of a precise value for $S_Z(2)$ in the initial singlet state of the combined system already demonstrates the incompleteness of the quantum-mechanical description, since this singlet state is certainly not an eigenstate of $S_z$ (2).[3]

It is very important for our subsequent discussion to notice that the EPR argument involves a locality assumption; that there was no change brought about in any element of reality associated with particle 2 by measurements performed on particle 1. Let us analyse this a little further in terms of possible interpretations of the uncertainty relations applied to noncommuting observables[4] such as spin-components projected along different directions.

Quite generally let us distinguish three possible answers to the question 'What value does an observable (call it $Q$) have when the state of the system is not an eigenstate of that observable[5]?'

(A) $Q$ has a sharp but unknown value.
(B) $Q$ has an unsharp or 'fuzzy' value.
(C) $Q$ has an undefined or 'meaningless' value.

With regard to (A) we may allow[6] that the unknown value is also unknowable, not in the sense that we cannot measure it and hence discover the value, but in the sense that it is not possible to know the value of $Q$ and retain also knowledge of the value of some noncommuting observable $Q'$ whose value is 'sharp' in the quantum-mechanical state in question. By 'sharp' here we mean that the state of the system is an eigenstate of $Q'$ so we can predict (with probability one if not with certainty) what the result of measuring $Q'$ will be. In the case of $Q$, however, we cannot make such predictions. Measurement of $Q$ will reveal different values in identically repeated experiments with probabilities (long-run frequencies) that are specifiable by the QM formalism.

With regard to (B) the terminology, although current in some elementary text-books on quantum mechanics, is really too vague. What is intended here is that $Q$ does not possess a *value* at all. What the system does in reality possess is a propensity[7] or potentiality[8] to produce various possible results on measurement. The propensity manifests itself in relative frequencies for various possible outcomes in a repeatable experimental arrangement (for measuring $Q$). Another way of representing position (B) is the idea of latency due to Margenau[9].

Turning to position (C) this is essentially the view taken by Bohr and is based on his complementarity interpretation of QM. For Bohr the definibility of an observable is not grounded in its measurability – it is just the other way about. The grounds for definibility are, in general mutually exclusive, experimental arrangements for displaying quantum-mechanical 'phenomena', which for Bohr involve in an essential way the specification of the experimental arrangements described in the unambiguous language of classical physics.

In summary we may say that the three views of the uncertainty relations we have indicated correspond to three general approaches to understanding QM:

(A) QM is not really mysterious at all – it is just a glorified statistical mechanics.

(B) QM is mysterious in that new concepts like potentiality or latency have to be introduced.

(C) QM is mysterious not in the sense of requiring new concepts but in the sense of recognising limitations on the applicability of the familiar concepts of classical physics, which may not be *definable* in certain contexts.

Again our three views are associated with three corresponding views of what measurement achieves:

(A) A pre-existing value is revealed.

(B) A potential result is actualized [10].

(C) An undefined value becomes defined, a meaningless quantity becomes meaningful.

After this brief digression let us return to the EPR argument. If we adopt position (A) we would have to allow at once the incompleteness of QM – the EPR argument for this would be redundant. If we adopt position (B) the EPR argument involves the following locality assumption which we distinguish as $\text{Locality}_1$:

> *$\text{Locality}_1$*: An unsharp value for an observable cannot be changed into a sharp value by measurements performed 'at a distance'.

The locution 'at a distance' in this definition may be taken in two possible senses which we distinguish as Bell locality and Einstein locality. For Bell locality 'at a distance' means 'in the absence of causal influences recognized by current physical theories' while for Einstein locality it means 'at a space-like separation between the space-time location of the system where the change from unsharp to sharp value of the observable is effected and the measuring apparatus producing this change'. If we accept provisionally that SR implies Einstein locality then we are claiming in the Einstein version of $\text{Locality}_1$ not just that no known physical causal influence is at work, but that no possible causal influence consistent with the constraints of SR could be effective in inducing the unsharp to sharp transition.

The EPR argument can now be represente schematically as

$$(\text{QM}) \wedge (\text{Locality}_1) \longrightarrow (\text{Incompleteness}),$$

or equivalently

$$(\text{QM}) \longrightarrow \sim (\text{Locality}_1) \vee (\text{Incompleteness}).$$

So we now have a dilemma. Either we must deny Locality$_1$ (even in the weak Einstein sense, since the measurements on the two particles can clearly be performed at a space-like separation), or QM is an incomplete theory in a sense which compels us to *give up* position (B) and turn to position (A).

EPR chose the second horn of the dilemma. We shall see below that this does not avoid nonlocality (of a different sort), but let us for a moment consider Bohr's response to the EPR argument (Bohr, 1935). Bohr effectively claimed that what was involved in the EPR argument was not Locality$_1$ but

> *Locality*$_2$: A previously undefined value for an observable cannot be defined by measurements performed 'at a distance'.

This principle could indeed be denied, and hence the argument for incompleteness circumvented, since no *physical* effect was involved. This move to block the EPR argument only makes sense if we adopt position (C) above in respect of values of observables not in an eigenstate, and hence depends on our accepting the definability criteria associated with Bohr's complementarity philosophy.

In the next section we shall revert to the second horn of the Einstein dilemma. Accepting that QM is incomplete let us attempt to complete the theory by associating sharp values at all times with all observables and see where this leads to.

## 3. THE BELL NONLOCALITY ARGUMENT

If QM is indeed an incomplete theory this suggests the programme of trying to complete it by ascribing simultaneous values to all observables including incompatible ones, that is to say to embed the partial or incomplete description afforded by QM in a more comprehensive theory which introduces a more refined specification of the state of a microsystem than that envisaged by the usual quantum-mechanical specification associated with rays in an appropriate Hilbert space. One way of doing this is to treat the QM observables as random variables defined on an underlying space of hidden variables and to define on this space a probability measure which will induce in these random variables the probability distributions demanded by the statistical algorithm of QM. The history of such attempts is well known. With suitable restrictions imposed different sorts of no-hidden-variable proof result, but when these restrictions are relaxed the hidden variable programme again becomes viable [11]. However in his (1964) Bell claimed that any hidden variable reconstruction of QM would involve a violation of a locality principle which we shall call

> *Locality*$_3$: A sharp value for an observable cannot be changed into another sharp value by altering the setting of a remote piece of apparatus.

Again we can distinguish an Einstein version as opposed to the stronger Bell version of this principle.

Various improvements were subsequently made in the proof of Bell's theorem [12], but until the work of Eberhard (1977), generalising the earlier suggestion of Stapp (1971), these all assumed a 'hidden variables' framework of one sort or another. Eberhard's proof of Bell's theorem does not make *this* assumption, but unfortunately was couched in a language of measurement results rather than possessed values of observables and was implicitly interpretated as demonstrating a violation of

> *Locality*$_4$: A macroscopic object cannot have its classical state changed by altering the setting of a remote piece of apparatus.

As we shall discuss in the next section this result of Eberhard's is in general false but we shall adopt his mathematics [13] *with a different interpretation* to demonstrate a violation of Locality$_3$.

Consider again Bohm's version of the EPR argument. Two spin-$\frac{1}{2}$ particles emerge from a source $S$ in a singlet spin state and move in opposite directions towards two spin-meters A and B which can measure the spin-projection of either particle along any specified direction. We shall consider two directions or 'settings' for each spin-meter, viz. $a$ and $a'$ for A and $b$ and $b'$ for B. For the $n$th pair of particles emitted from the source denote by $a_n$ the spin-component of the A-particle (i.e., the particle travelling towards spin-meter A) projected in the direction $a$ in units of $\hbar/2$ when the A-meter is set parallel to $a$. Similarly for $a'_n$, $b_n$ and $b'_n$ in an obvious notation. Clearly QM predicts that insofar as measurements by the spin-meters merely reveal these values, then these values are restricted always to be $\pm 1$.

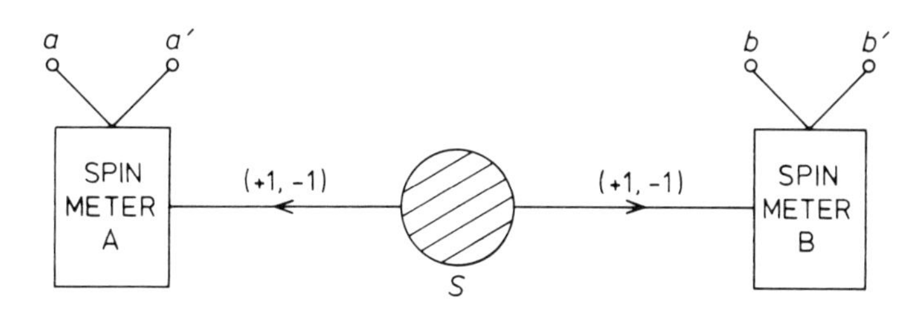


Fig. 1.

The situation is sketched in Figure 1, where the two settings of each spin-meter are indicated by the possible positions labelled $a$, $a'$ and $b$, $b'$ of a 'knob' or joy-stick attached to the corresponding meter.

Now form the expression (following Eberhard)

$$\gamma_n = a_n b_n + a_n b'_n + a'_n b_n - a'_n b'_n.$$

$\gamma_n$ clearly has integral values which can at most lie between −4 and +4. The trick here is that the value of the fourth term (with the minus sign) is the product of the first three. This restricts the value of $\gamma_n$ to $\pm 2$, as the following simple argument confirms. Write

$$\gamma_n = a_n(b_n + b'_n) + a'_n(b_n - b'_n).$$

Now $b_n$ and $b'_n$ must have either the same sign or opposite sign. In either case only one term is non-vanishing and its value is clearly $\pm 2$.

Now consider $N$ events and form

$$\left|\frac{1}{N}\sum_{n=1}^{N}\gamma_n\right| = \left|\frac{1}{N}\sum_{n=1}^{N} a_n b_n + \frac{1}{N}\sum_{n=1}^{N} a_n b'_n + \frac{1}{N}\sum_{n=1}^{N} a'_n b_n - \frac{1}{N}\sum_{n=1}^{N} a'_n b'_n\right| \leqslant 2.$$

If we define correlation coefficients $c(a,b) = \lim_{N\to\infty}\Sigma_{n=1}^{N} a_n b_n$, etc. then the above inequality yields

$$|c(a, b) + c(a, b') + c(a', b) - c(a', b')| \leqslant 2.$$

This is one form of the so-called Bell inequality which is readily shown to be violated by QM for suitable choice of the directions $a$, $a'$, $b$ and $b'$. Thus for QM we can easily evaluate

$$c(a, b) = -\cos\theta_{ab} \simeq -1 + \tfrac{1}{2}\theta_{ab}^2 - \ldots,$$

where $\theta_{ab}$ denotes the angle between the directions $a$ and $b$. Notice how as we move away from a situation of perfect anticorrelation ($\theta_{ab} = 0$) the QM correlation function 'hangs on' to the anticorrelation more tightly than is the case in classical physics where typically decorrelation is in proportion to $\theta$ rather than $\theta^2$.[14] Choose the directions $a, a', b, b'$ to be coplanar and take $a$ parallel to $b$ and $\theta_{ab'} = \theta_{a'b} = \phi$, say, so $\theta_{a'b'} = 2\phi$. Then Bell's inequality reduces to

$$F(\phi) =_{\text{Df}} |1 + 2\cos\phi - \cos 2\phi| \leqslant 2.$$

In Figure 2 below we show $F(\phi)$ plotted as a function of $\phi$ in the range $0 < \phi < 180°$. The Bell inequality is violated for all values of $\phi$ between 0 and 90°, the maximum value for $F(\phi)$ of $2\frac{1}{2}$ being attained for $\phi = 60°$.[15]

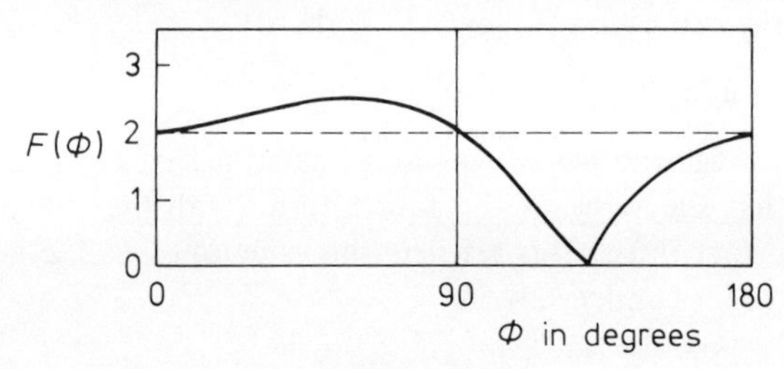


Fig. 2.

The essential ingredient that has gone into the proof of Bell's theorem is the assumption of Locality$_3$. For example we have assumed that the value $a_n$ is the same whether we are measuring $b_n$ or $b'_n$, that the change in setting of the knob on the spin-meter $B$ from $b$ to $b'$ does not affect the value $a_n$ (which is 'discovered' by the spin-meter $A$ with its knob set in the direction $a$). Note, however, that our definition of $a_n$ does allow for a dependence of the spin-projection of the A-particle parallel to $a$ on the setting of the spin-meter A.

There are four separate correlation experiments involved in testing the Bell inequality in the form in which we have presented it. These involve combining setting $a$ with $b$, $a$ with $b'$, $a'$ with $b$ and $a'$ with $b'$ respectively. The four experiments are mutually exclusive in the sense that each knob can have only one setting for any given experiment so we are not allowing the possibility of measuring $a_n$, $b_n$, $a'_n$, $b'_n$ simultaneously (cf. Note 6).

Nevertheless we are assuming that $a_n$, $b_n$, $a'_n$, $b'_n$ all have definite values which can be measured simultaneously in pairs; $a_n$ with $b_n$, $a_n$ with $b'_n$, $a'_n$ with $b_n$ and $a'_n$ with $b'_n$.

We illustrate what is going on by considering the table of values for $a_n$, $a'_n$, $b_n$ and $b'_n$ (see Figure 3), in which the four correlation experiments are distinguished as I, II, III, IV.

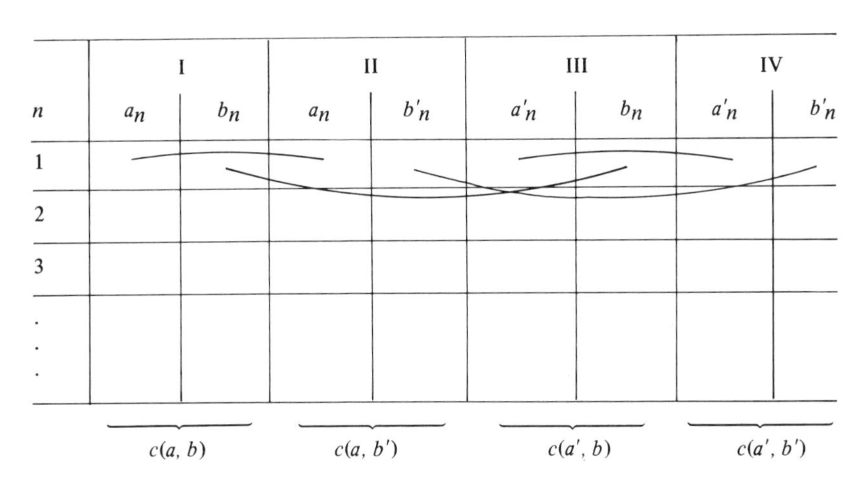


Fig. 3.

Each of the four values occur twice in each row of the figure, since it figures in two correlation experiments. The fact that each of the two occurrences has the same numerical value (indicated by the 'ties' in Figure 3) we term the Matching Condition, which essentially incorporates the assumption of Locality$_3$. Each pair of columns for a given correlation experiment enables one to compute a correlation coefficient with respect to (in general counterfactually) *possessed* values using *all* values of $n$. But to carry out the correlation experiments we must perform a *place-selection* on this sequence of possessed values which tell us which values of $n$ are to be the subject of which measurement procedure (I, II, III, or IV). In the preceding discussion we have tacitly made two additional assumptions:

(1) Limiting frequencies computed under these place-selections have the same value as those computed with all values for $n$.

(2) The correlation coefficients evaluated with respect to *measured* values are the same as those evaluated with respect to possessed values.

The first assumption is merely that each sequence of $a_n$'s, $b_n$'s, etc. is a random sequence.

The second assumption has been challenged for example by Fine (1979 and 1980) but the fact, if it is a fact, that we cannot probe possessed correlations by measured correlations, suggests a remarkable conspiracy on the part of Nature to conceal the former. In what follows we shall ignore *conspiracy* models (called by Fine in his (1980) synchronization models) for explaining the results of the correlation experiments. Notice that we do not assume that the $a_n$'s and $a'_n$'s for example have a well-defined joint probability distribution, the correlation functions actually used in deriving the Bell inequality referring always to compatible (commuting) observables. In particular we do not assume $1/N\Sigma_{n=1}^{N} a_n a'_n$ has any well-defined limit as $N \longrightarrow \infty$. This is an important respect in which the above proof is superior to the original 'hidden variables' proof of Bell [16]. Furthermore, we do not assume that $a_n$ and $a'_n$ can be measured simultaneously contrary to the view expressed by Brody and de la Peña-Auerbach (1979) (see also Brody, 1980).

So if the Bell inequality is violated by the experimental measurements of the appropriate correlation coefficients we may conclude that $\text{Locality}_3$ is violated [17]. The experimental evidence now seems conclusive that in the Bell sense this is the case [18] although the more delicate question of violating $\text{Locality}_3$ in the Einstein sense still awaits the results of experiments in which the settings of the measuring devices are changed after the particles are actually emitted from the source[19].

In this way we have blocked the second horn of the EPR dilemma. The first horn led to violation of $\text{Locality}_1$, the second horn is now seen to lead to violation of $\text{Locality}_3$.

## 4. THE STAPP-EBERHARD APPROACH

We now want to turn to the question of whether the experimental violation of the Bell inequality can be interpreted as a violation of $\text{Locality}_4$ in the terminology introduced at the beginning of the previous section. The idea of formulating the Bell argument entirely in terms of the actual or possible responses of macroscopic measuring apparatus was introduced by Stapp (1971), and developed in the work of Eberhard (1977) already referred to. The Stapp–Eberhard approach seeks to remove from the argument any assumptions about the properties of micro-systems such as the views we distinguished as (A), (B) and (C) in Section 2. Thus in the mathematics of the Eberhard proof of Bell's inequality as presented in Section 3 above, we now interpret the quantities $a_n$, $a'_n$, $b_n$ and $b'_n$ not as possessed values for the micro-systems, but as the response of the macroscopic spin-meters when set to measure these quantities. However the four correlation experiments cannot be performed simultaneously, so Stapp and Eberhard introduce (effectively) a

> *Principle of Local Counterfactual Definiteness* (*PLCD*): The result of an experiment which *could* be performed on a micro-system has a definite result which does not depend on the setting of a remote piece of apparatus.

This principle is apparently seen by Stapp and Eberhard as licensed by $\text{Locality}_4$. Now PLCD can be used to justify the Matching Condition involved in the proof of the Bell inequality since the value $a_n$ for example is the same for experiments I and II, independently of whether the setting of spin-meter B is along $b$ or $b'$. Since the Bell inequality is violated by experiment we must infer that $\text{Locality}_4$ is violated.

But let us look at PLCD more closely. In a deterministic situation in which individual measurement results are predictable on the basis of a more complete theory than QM, PLCD is quite innocuous. Indeed, in the previous section our definition of $a_n$, $b_n$, etc. involved a harmless counterfactual element. But Stapp and Eberhard are specifically embracing the possibility of essentially indeterministic measurement outcomes, as in the position we labelled (B) in Section 2. However in this case PLCD is highly suspect.

Let us take a simple example to bring out the point at issue. At atom of radium is situated at one end of a table. I stand at the other end of the table and at time $t_1$ I raise my hand. Let us suppose that at time $t_2$, shall we say one second later than $t_1$, the atom of radium decays (emits an $\alpha$-particle). Now let us ask the question: 'If I had not raised my hand at time $t_1$ would the atom still have decayed at time $t_2$?' the PLCD says the answer to this question is 'Yes'. But let us suppose that the fact of whether the atom of radium decays at $t_2$ is indeterministic, that is to say there is no refinement of our description of the state of the universe prior to $t_2$ on the basis of which the decay could have been predicted. All that we can predict is the probability of a decay at $t_2$[20]. We have now to analyse the truth condition for the conterfactual $\phi \mathrel{\Box\!\!\rightarrow} \psi$ where $\phi$ denotes the condition that I do not raise my hand and $\psi$ the state of affairs that the atom decays at time $t_2$. Let us try a possible worlds analysis. Let the world in which I raise my hand at $t_1$ and the atom decays at $t_2$ be denoted by $W_i$. Let possible worlds $W_j$

for variable $j$ be ordered in respect of 'nearness' to $W_i$. If $W_l$ is nearer to $W_i$ than $W_k$ I write $W_l \underset{i}{<} W_k$. Then I analyse $\phi \;\Box\!\!\rightarrow \psi$ as

$$\exists W_k \, [\exists W_l((W_l \underset{i}{<} W_k) \wedge W_l(\phi))$$
$$\wedge \, \forall W_j((W_j \underset{i}{<} W_k) \rightarrow (W_j(\phi) \rightarrow W_j(\psi)))]$$

where $W_p(\phi)$ signifies that $\phi$ is true in $W_p$[21]. We are assuming then $\sim W_i(\phi)$.

> In words: there is a world sufficiently close to $W_i$ such that there exist some worlds closer to $W_i$ in which $\phi$ holds and such that for *any* such world $\psi$ holds.

Now the specification of sufficiently close must refer to states of the world up to but not including $t_2$, i.e., we must not include the fact of $\psi$ obtaining at $t_2$ as part of the specification of a sufficiently close world.[22] Informally we consider a world which differs from our actual world just in the fact that $\phi$ obtains in the alternative world but not in the actual world, everything else including the laws of nature are the same and we let the world run on to the instant before $t_2$ and ask 'Must now $\psi$ occur at $t_2$?'. But if the occurrence of $\psi$ is essentially probabilistic then there is no necessity for $\psi$ to occur at $t_2$ in the alternative world. Indeed if the actual world were 'run' over again with $\phi$ *not* obtaining at $t_1$, we still could not infer that $\psi$ *must* occur at $t_2$. This is not paradoxical. It is just what we mean by saying the occurrence of $\psi$ is indeterministic (essentially probabilistic). We just cannot refine the description of the world prior to $t_2$ to force the occurrence of $\psi$ at $t_2$. If such a refinement were possible the occurrence of $\psi$ would be deterministic, not indeterministic.

To return to our original example, the fact that lowering my hand at $t_1$ *allows* the atom *not* to decay at $t_2$ has nothing to do with a violation of locality – it just involves the recognition of what is meant by the claim that the decay of a radium atom is indeterministic. So what is wrong with the PLCD is the meaning attached to 'definite'. The outcome of an essentially indeterministic situation is definite in the sense that on a *particular* occasion, in a *particular* world, whatever does happen is the determinate outcome, but it is not definite in the sense that it is both determinate and *determined* by any possible specification of that world prior to the occurrence of that particular outcome[23].

The upshot of this discussion is that on view (B) of the quantum world the PLCD cannot be invoked as a valid principle licensed by an appeal to

$\text{Locality}_4$. So the Matching Condition in the Eberhard proof does not follow in any way from $\text{Locality}_4$ and the fact that the result derived from the Matching Condition, viz. the Bell inequality, is violated by experiment does not imply that $\text{Locality}_4$ is violated. All this depends on adopting view (B). On view (A) violation of $\text{Locality}_3$ clearly implies violation of $\text{Locality}_4$. It is the generality claimed for Eberhard's proof of $\text{nonlocality}_4$ that is being challenged.

There is an alternative way that Eberhard refers to in his (1977) for expressing his results which does not employ counterfactuals. The idea is to record in a table such as that illustrated above not the results of four correlation experiments which could have been performed, although not simultaneously, but the results of four correlation experiments which are all actually performed. In other words each pair of columns records a sequence of measurement made with the appropriate pair of knob settings. Having obtained four such correlation sequences they are written down side by side to form the complete 8-column table, but each row now refers to four *different* particle pairs emitted by the source. So clearly in general the Matching Condition will not hold for any particular row. But, by chance, it may. So now perform a place-selection on the columns which consists in selecting those rows for which the Matching Condition *does* hold. If we calculate now correlation coefficients $c(a, b)$, etc., using only these place-selected rows of the table, then Eberhard points out that the results cannot in general agree with the correlation coefficients calculated with the whole table. This is perfectly true since the former, *by construction*, satisfy the Bell inequality, whereas the latter, in general, violate it. But all this has nothing to do with $\text{Locality}_4$. It just demonstrates the familiar fact that place-selections in a random sequence, made in the light of the actual outcomes, can change the limiting frequencies for these outcomes.

## 5. STATISTICAL NONLOCALITY

There is still another sense of locality that enters into the discussion of the EPR paradox.

> *$\text{Locality}_5$*: The statistics (relative frequencies) of measurement results of a quantum-mechanical observable cannot be altered by performing measurements 'at a distance'.

It is important to realise that $\text{Locality}_5$ is not violated in an EPR type of measurement.

To see this let us suppose we introduce a measuring device which we label as system 3 and use it to measure the value of $S_z(1)$ in the state $\Psi_{\text{singlet}}$. Let the initial state of the apparatus be denoted by $W_0(3)$ and the final state be denoted by $W_+(3)$ if $S_z(1)$ has the value $+\frac{1}{2}\hbar$ and $W_-(3)$ if $S_z(1)$ has the value $-\frac{1}{2}\hbar$. Then, assuming ideal measurement, the state of the whole system 1, 2 and 3 goes from $\Psi_i = 1/\sqrt{2}(\alpha(1)\beta(2) - \beta(1)\alpha(2))W_0(3)$ before the measurement to $\Psi_f = 1/\sqrt{2}(\alpha(1)\beta(2)W_+(3) - \beta(1)\alpha(2)W_-(3))$ after the measurement. In terms of density operators, it is well known that $P_{\Psi_i}$ behaves like $\frac{1}{2}P_{\alpha(2)} + \frac{1}{2}P_{\beta(2)}$ in respect of measurements of an observable pertaining to particle 2 only. $P_{\Psi_i}$ describes a so-called improper mixture in the terminology of d'Espagnat [24]. But $P_{\Psi_f}$ is also an improper mixture and again behaves like $\frac{1}{2}P_{\alpha(2)} + \frac{1}{2}P_{\beta(2)}$ in respect of measurements of an observable pertaining to particle 2 only [25]. So the statistics of measurement results for any observable pertaining only to particle 2 is unaffected by 'hooking on' the apparatus for measuring the $Z$ spin-component of particle 1. We are referring here of course to the non-selective stage of measurement. If we select a subensemble of particle 2's with $S_z(2) = +\frac{1}{2}\hbar$ say, this will be described by the density operator $P_{\alpha(2)}$, which of course gives different statistics from $\frac{1}{2}P_{\alpha(2)} + \frac{1}{2}P_{\beta(2)}$. But the selection is made at the wrong location to produce any 'instantaneous' statistical effects 'at a distance'.

Thus suppose we are measuring a sequence of values for $S_z(2)$ on successive particles emitted by the source. The sequence might be $+ - - + - + + -$ ..., where the limiting frequencies of + and − is $\frac{1}{2}$. Now we perform measurements simultaneously on particle 1. This enables us to 'tag' each particle 2 as either + or − in our abbreviated notation for values of $S_z(2) = \pm\frac{1}{2}\hbar$. But the tagging information is in the wrong place to change the statistics at the location of particle 2. To do this we would have to transmit the tagging information from location 1 to location 2, with instructions, for example, to insert an absorbing screen every time *a*−particle is approaching the spin-meter for particle 2, and to remove it every time *a*+particle is approaching. In this way we would change the above sequence to $+ + + + +$ ..., but to effect this change, we have to transmit information from location 1 to location 2; we cannot do it simply by 'hooking on' the apparatus to measure $S_z(1)$.

To give a simple example to illustrate the problem. When I lecture in Oxford the audience *there* learn instantaneously that my room in London is empty, but to produce a physical change in London, for example to prevent students knocking on my door, the information that I have arrived in Oxford must be transmitted *back* to London. In brief the 'Bell' telephone for transmitting information instantaneously between two remote locations does not and cannot work [26]. In this respect the EPR situation is quite different from what would obtain if operators referring to particles 1 and 2 failed to commute. In such a case $\text{Locality}_5$ would clearly be violated. But the nonlocality demonstrated by EPR is subtler than this.

The fact that no statistical effects get transmitted at a distance means that the EPR problems do not arise in an ensemble or statistical interpretation of QM. It is only in the context of an attempt to inpute states to *individual* systems that the difficulties are manifested.

## 6. PROPENSITIES AND PEACEFUL COEXISTENCE

Shimony in his (1978) has considered the question of whether the violation of $\text{Locality}_1$ should be considered a serious matter in the light of relativistic constraints. After all, Shimony remarks, the 'event' which consists in actualizing a possibility is a bit mysterious anyway and one might want to argue that SR applies to 'concrete' events like violations of $\text{Locality}_3$ or even more certainly $\text{Locality}_4$. So perhaps we can have 'peaceful coexistence' between SR with its usual interpretation as incorporating a First Signal Principle (FSP) and violations of $\text{Locality}_1$ if this is how we choose to interpret the Bell inequality experiments. We can also retain realism, but now a realism about propensities rather than possessed values.

However, I now want to consider and to reject an argument that violation of $\text{Locality}_1$ is not really an instance of nonlocality at all and hence that 'peaceful coexistence' is quite unproblematic [27]. The argument goes like this. Connecting up the spin-meter for particle 1 in no way affects the statistical properties of particle 2 considered by itself – it is only the relational properties between 1 and 2 that get altered by the measurement, as we explained in the previous section. But relational properties of particle 2 with respect to particle 1 can be changed by just changing one of the relata, viz. particle 1, and this is what the spin-meter does, there is no nonlocal effect of the spin-meter that measures $S_z(1)$ on particle 2 at all. But this argument makes it look as though the 'fuzziness' in properties of particle 2 in respect of the $Z$-component of its spin are not inherent in particle 2 but arise from its relations to something which does have fuzzy properties, viz. particle 1, and this latter fuzziness can be removed locally by measurements performed on particle 1. But this cannot be right because equally we can remove all fuzziness including relational effects by measurements performed just on particle 2. The situation is quite symmetrical between the two particles.

Any attempt to impute relational fuzziness to just one of the particles must fail, since what is true of one particle must also be true of the other. So we just cannot give a 'local' explanation of how relational fuzziness gets removed by measurements performed on one or other particle. Relational fuzziness is inherently a property of both particles and its removal is necessarily a nonlocal effect, a violation of $\text{Locality}_1$.

If we do not take Shimony's line of peaceful coexistence and prefer not to follow Bohr's line of using $\text{Locality}_2$, there is still the possibility that SR when properly understood, does not prohibit violations of locality at all. This is the problem we shall look at next.

## 7. SUPERLUMINAL INTERACTIONS IN SPECIAL RELATIVITY

It is a well-known paradox in the literature on SR that, while philosophers like Reichenbach and Grünbaum investigating the foundations of the subject have concluded that the possibility of physical effects being propagated with velocities exceeding in magnitude the velocity of light in vacuo ($c$) is inconsistent with SR [28], nevertheless physicists for the last twenty years have been predicting the possible occurrence in nature of so-called tachyons (particles moving faster than $c$) *on the basis of SR* [29]. Of course the resolution of this paradox lies in distinguishing what exactly is meant by 'SR'. The physicist's understanding of SR is that it is encompassed by the invariance principle (IP) which asserts that all the laws of physics assume the same mathematical form when referred to any inertial frame of reference, the connection between space-time coordinates in any two inertial frames $K$ and $K'$ being given by the familiar Lorentz transformation

$$(1)\qquad \begin{aligned} x' &= 1/\sqrt{1-v^2/c^2}\;, \quad (x - vt) \\ y' &= y \\ z' &= z \\ t' &= 1/\sqrt{1-v^2/c^2} \cdot (t - vx/c^2) \end{aligned}$$

where we consider the case where $K'$ moves with uniform velocity [30] $v$ along the positive $X$-axis of the $K$-frame.

Now there is a purported argument that IP → FSP where FSP is the First Signal Principle, that there is an upper limit, viz. $c$, to the velocity with which a signal can propagate relative to an inertial reference frame. Thus Einstein in his (1905) noted that an infinite amount of energy would be required in SR to accelerate a particle of finite rest-mass to the velocity $c$, and hence concluded 'velocities greater than that of light have ... no possibility of existence"[31] and indeed "the velocity of light in our theory plays the part physically of an infinitely great velocity"[32]. The fallacy here lies in the fact that SR provides no prohibition on describing particles that move already with velocities greater than $c$, only that $c$ plays the role of a velocity 'barrier' that cannot be crossed from below or indeed from above. So we have a three-fold classification with the now standard nomenclature

*Bradyons*: particles which move always with velocities less than $c$;
*Photons*: particles which move always with velocities equal to $c$;
*Tachyons*: particles which move always with velocities greater than $c$.

Tachyons have some rather surprising dynamical properties. The energy $E$ of a tachyon which moves along the positive $X$-axis with a velocity $U_x > c$ is given by

$$(2) \qquad E = \mu c^2 / \sqrt{U_x^2/c^2 - 1}\ ,$$

where $\mu$ is a mass-parameter (not of course rest-mass, tachyons remember cannot be brought to rest).

As $U_x$ increases $E$ decreases and actually tends to zero as $U_x$ tends to infinity.

Relative to a new frame $K'$ moving with velocity $v$ relative to $K$ along the positive $X$-axis[33], the tachyon now has energy given by

$$(3) \qquad E' = E/\sqrt{1 - v^2/c^2} \cdot (1 - vU_x/c^2).$$

If $v > c^2/U_x$ $E'$ becomes negative, but for the time-ordering of two events on the trajectory of the tachyon, we also have a reversal of sign at the *same* critical frame velocity for $K'$. Thus it is easy to show that if we label the events as 1 and 2,

$$(4) \qquad t_2' - t_1' = \frac{t_2 - t_1}{\sqrt{1 - v^2/c^2}} \cdot (1 - vU_x/c^2),$$

which again changes sign for $v > c^2/U_x$.

This leads to the so-called Reinterpretation Principle (RIP), a tachyon of negative energy moving backwards in time is reinterpretated as an anti-tachyon of positive energy moving forwards in time[34]. In this way, it is argued, backward causation effects associated with the apparent reversal of the time-ordering of events on a tachyon trajectory relative to an appropriately moving reference frame can be eliminated. Thus in the formal[35] theory a tachyon could be emitted at $A$ at time $t_1$ and move with positive energy to location $B$ at time $t_2$ and then propagate back to $B$ with negative energy arriving there at a time $t_3$ prior to $t_1$ (see Figure 4).

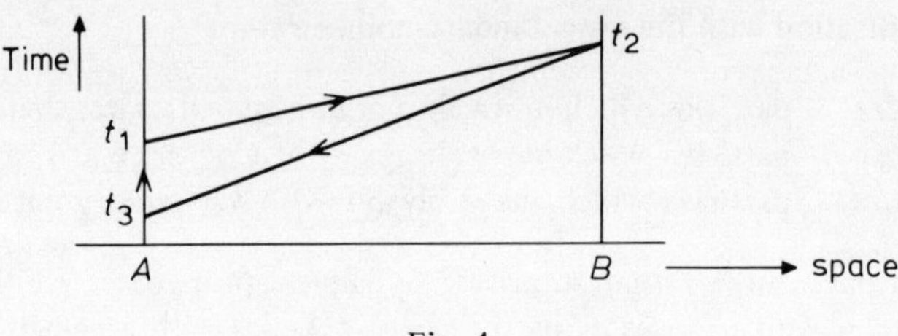


Fig. 4.

If such a device could be used to transmit a message such as 'If and only if this message is received do not emit a tachyon at time $t_1$', a logical paradox at once arises. If a tachyon is emitted at $t_1$ we conclude it will not be emitted. If it is not emitted at $t_1$, then it will be emitted. With the application of RIP Figure 4 looks like Figure 5.

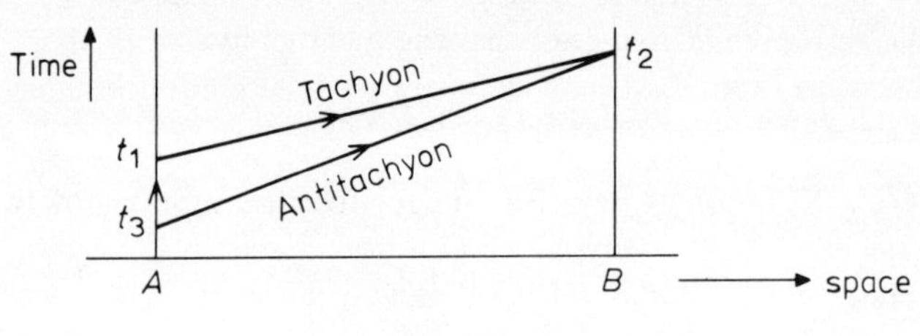


Fig. 5.

A tachyon and an antitachyon are emitted at $A$ at times $t_1$ and $t_3$ respectively and annihilate one another when they meet at location $B$ at time $t_2$. The 'causal loop' in Figure 4 has been eliminated by reversing the direction of the antitachyon's causal efficacy, it is now regarded as transmitting an effect from $A$ to $B$ rather than from $B$ to $A$. All this makes sense in terms of energy balance, since absorbing a particle of negative energy is clearly equivalent, from the energy balance point of view, to emitting a particle of positive energy.

The idea that the possibility of tachyons could be ruled out on the basis of a causality paradox goes back to Tolman (1917). There has been vigorous discussion in the literature as to whether RIP enables one to solve such paradoxes in an acceptable way. New versions of the causal loop paradox were produced by Newton (1967), Rolnik (1969), Benford *et al.* (1970) and Pirani (1970). In the Pirani version, for example, four observers *A, B, C, D* are involved in sending tachyon signals in a causal loop that begins with emission of a tachyon by *A* which is received by *B*, who transmits a signal to *C* and so on back to *A*. The four observers are in appropriate relative motion such that, relative to the rest-frame of each observer, he sees one (positive energy) tachyon as being *absorbed* and another *emitted* while the initial observer *A*, after emitting a tachyon that sets the causal loop in operation, sees the absorbtion of the final (positive energy) tachyon at an earlier time. In a proposed resolution of this paradox Parmentola and Yee (1971) stressed that Pirani had failed to use *one* reference frame from which to describe the whole sequence of events. When only one frame was used they showed that the causal loop could not be established. What emerges from these discussions is that the notion of absorbing or emitting a tachyon is not an absolute one, an event described as emission in one frame of reference is described as absorbtion (of an antitachyon) in another and vice versa. Hence the idea of sending a tachyonic signal via a chain of emission and absorbtion events becomes very problematic, and this is what the causal loop paradoxes depend on [36]. However, this does not rule out the possibility that tachyonic interaction might sustain a *symmetric* relation of causal connectibility between events at space-like separation, although an asymmetric cause-effect relation could not be established. Now, in the philosophical discussion of the foundations of SR given by Grünbaum in his (1973), a notion of causal connectibility is used to ground a notion of temporal betweenness, which in turn underpins the admissibility of the clock-synchronization procedures needed to sustain the Invariance Principle in SR. It is by following this sequence of arguments that Grünbaum claims to demonstrate an inconsistency between 'tachyonic' causal connectibility and the IP in the sense that

$$\text{PIP} \longrightarrow \text{FSP},$$

where PIP denotes 'Philosophically grounded Invariance Principle'. We are not here deriving FSP illicitly from IP as in the Einstein argument discussed above, but from the first 'P' in PIP if I may so express myself, since FSP is the 'philosophical ground' for licencing IP on the Grünbaum analysis. That is to say IP, when properly understood, incorporates FSP as a presupposition.

But FSP, says Grünbaum, is contradicted by 'tachyonic' interactions. Hence tachyons are inconsistent with SR, now understood as embracing PIP, not just IP as in the physicist's analysis we have been discussing so far. In the next section I will attempt to evaluate this line of attack on the possibility of superluminal interactions in SR.

## 8. THE CAUSAL THEORY OF TIME AND THE INVARIANCE PRINCIPLE

I shall begin by recapitulating briefly the way in which Reichenbach and Grünbaum employ the notion of causal connectibility to justify the clock-synchronization procedures implicit in formulating the Invariance Principle in the usual physicist's account of SR. I refer in particular to the analysis given in Grünbaum (1973) where a sufficient condition for the causal connectibility of two events $e_1$ and $e_2$ is taken to be the existence of a continuous genidentical set of events of which $e_1$ and $e_2$ are both members. The notion of a genidentical set of events as one in which all the events belong to the space-time history of a (reidentifiable) material particle is taken as primitive, as also is the notion of a continuous (or $k$-connected) genidentical set as one with no 'gaps' in it [37]. Consider two spatial locations $A$ and $B$. A signal originates from $A$, event $e_1$, propagates to $B$, its arrival there being event $e_3$, and then propagates back to $A$, event $e_2$, as sketched in Figure 6.

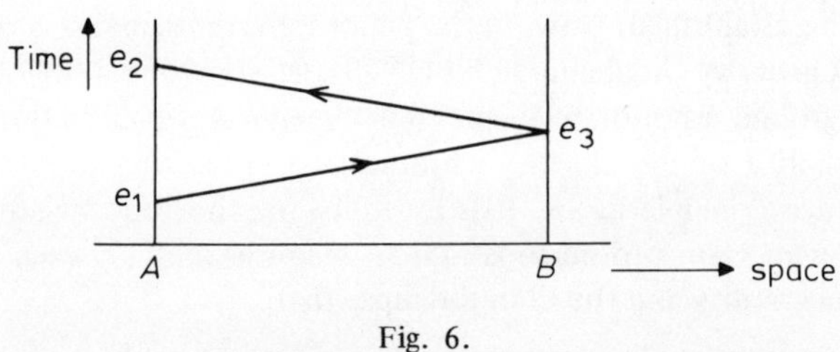


Fig. 6.

We now introduce the following definition: $e_3$ is *temporally between* $e_1$ and $e_2$ (we shall write this as $e_3 B e_1 e_2$) if and only if

(1) $e_1, e_2, e_3$ are members of a possible genidentical set $G$

and

(2) If $e_3$ is deleted from $G$ there is no continuous genidentical subset of $G$ that contains both $e_1$ and $e_2$.

Formally using $G$ as a variable ranging over all possible complete (and hence continuous) genidentical sets and $S$ as a variable ranging over all possible continuous (but not necessarily complete) genidentical sets, so $\forall S \exists G(S \subset G)$, then we have

$$e_3 B e_1 e_2 \quad \text{iff}$$
$$\exists G[G(e_1, e_2, e_3) \wedge \forall S(S \subset G - \{e_3\} \rightarrow \sim S(e_1, e_2))]$$

where $(e_1 \in G) \wedge (e_2 \in G) \wedge (e_3 \in G)$ is written $G(e_1, e_2, e_3)$ and similarly for $S(e_1, e_2)$.

We notice that the relation $B$ may fail to obtain for two quite different reasons. If condition (1) holds but not condition (2) in the above definition then we may say that $e_3$ is absolutely outside the temporal interval $e_1$, $e_2$ but if condition (1) does not hold then the relation $B$ fails to obtain in a more decisive sense that we may want to express by saying that a *betweenness* relation is not really now *applicable* because $e_3$ is in an absolute sense *simultaneous* with either or both of the events $e_1$ and $e_2$. To capture this idea of absolute or topological simultaneity between an event $e_3$ at location $B$ and any event, call it $e_4$, at location $A$, Grünbaum makes effectively the following definition

$$e_3 S_G e_4 \quad \text{iff} \quad \forall G(\sim G(e_3, e_4)).$$

We use the notation $S_G$ for this Grünbaum definition of simultaneity. The relations $B$ and $S_G$ now mesh in the following way

*The Meshing Condition*: $\sim \exists G(G(e_1, e_2, e_3))$ holds if and only if $e_1 S_G e_3$ or $e_2 S_G e_3$.

The Meshing Condition involves two claims

(1) $$\exists G(G(e_1, e_2, e_3)) \rightarrow \exists G(G(e_1, e_3)) \wedge \exists G(G(e_2, e_3))$$

and

(2) $$\exists G(G(e_1, e_3)) \wedge \exists G(G(e_2, e_3)) \rightarrow \exists G(G(e_1, e_2, e_3)).$$

The first part, being a logical truth, is not controversial. Provided we restrict our consideration to causal interactions effected by bradyons and photons the second part of the Meshing Condition is also clearly valid. We shall return later to discuss the problems posed for the Meshing Condition when tachyon interactions are taken into account.

We are now in a position to formulate the First Signal Principle FSP in a form appropriate to discussing the philosophical underpinning of IP. Using now $e_1$, $e_2$ and $e_4$ as variables ranging over events at location $A$ we assert

$$\exists e_1 \exists e_2 [e_3 B e_1 e_2 \wedge \forall e_4 (e_4 B e_1 e_2 \longrightarrow e_3 S_G e_4)].$$

Denote by $\bar{e}_1$, $\bar{e}_2$ the two events at location $A$ corresponding to the sending out and return of the 'First Signal', which arrives at $B$ as event $e_3$. Then, says Grünbaum, $e_3$ is topologically or absolutely simultaneous with any event $e_4$ at location $A$ which lies temporally between $\bar{e}_1$ and $\bar{e}_2$. So now we can re-set the zero of our standard clock at location $B$ so that its metrical reading $t_3$ has any value lying between the metrical readings $t_1$ and $t_2$ of the standard clock at $A$ corresponding to the events $\bar{e}_1$ and $\bar{e}_2$. In this way topological simultaneity as characterised by the FSP is translated into an ontological latitude or ambiguity in assignments of metrical simultaneity, which is summarized in the famous Reichenbach formula

$$(1) \qquad t_3 = t_1 + \epsilon(t_2 - t_1),$$

where $\epsilon$ is a parameter with *any* value between 0 and 1, whose actual value can be fixed only as a matter of *convention.*[38] Identifying the First Signal with the propagation of a light wave (photon) the parameter $\epsilon$ is easily seen to be related to the two one-way velocities of light $\vec{c}$ and $\overleftarrow{c}$ by

$$(2) \qquad \begin{aligned} \vec{c} &= \bar{c}/2\epsilon \\ \overleftarrow{c} &= \bar{c}/2(1-\epsilon), \end{aligned}$$

where $\bar{c}$ is the average two-way velocity of light.

The Einstein convention takes $\epsilon = \frac{1}{2}$, giving $\vec{c} = \overleftarrow{c} = \bar{c} = c$, say.

If we now consider a reference frame $K'$ moving with velocity $v$ along the positive $X$-axis, the relation between the new space-time coordinates $(x', t')$ and the coordinates $(x, t)$ in the frame $K$ (for simplicity we confine ourselves for the moment to two-dimensional space-time) is given by some linear relationship[39]

(3) $$x' = Ax + Bt,$$
$$t' = Ct + Dx.$$

For $x' = 0$, $x = vt$, so $B = -Av$ and if we define $m = -D/C$, then we obtain

(4) $$x' = A(x - vt)$$
$$t' = C(t - mx).$$

Now suppose a rod stationary in $K'$ but moving relative to $K$ with a velocity $v$ is contracted in length by a factor $F$, while a moving clock is similarly dilated in time interval by a factor $G$[40], then it follows immediately that

$$A = 1/F, \quad C = 1/G(1 - mv)$$[41]

so we now have

(5) $$x' = 1/F \cdot (x - vt)$$
$$t' = 1/G(1 - mv) \cdot (t - mx).$$

It is of course well known that if we adopt the Einstein convention in $K$ then the values of $F$ and $G$ are given by the relativistic formulae

(6) $$F = 1/G = \sqrt{1 - v^2/c^2}\,.$$

At this point we have included all the *physics* that is involved in SR. The significance of the parameter $m$ is clearly that the line $t = mx$ gives the locus of events simultaneous with the origin relative to the frame $K'$. That is to say $m$ is the slope of the line of simultaneity for $K'$ assessments as seen from $K$. The value of $m$ is fixed by the simultaneity convention ($\epsilon$-value) adopted in $K'$. In particular if we adopt the Einstein convention *also* in $K'$, then $m$ is easily seen to be given by[42]

(7) $$m = v/c^2.$$

Substituting in (5) and using (6) yields the usual form of the Lorentz transformations given as Equation (1) in Section 7.

The important question now is whether ontologically speaking we are *justified* in choosing the Einstein convention in $K'$.

*We define*: A choice of synchronization in $K'$ is said to be *bizarre* if it makes metrically simultaneous relative to $K'$ events which are not topologically simultaneous relative to $K$.

We now have the important

THEOREM. The choice of the Einstein convention in $K'$ is never bizarre.

The proof is best seen from Figure 7 below.

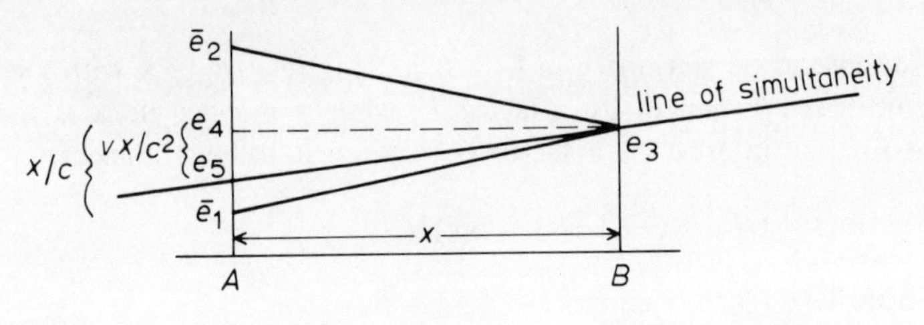


Fig. 7.

Relative to the event $e_3$ at $B$ the event at $A$ which is metrically simultaneous with $e_3$ relative to the Einstein convention in $K'$ is the event $e_5$ which precedes the event $e_4$ which is metrically simultaneous with $e_3$ according to the Einstein convention in $K$ by a time $vx/c^2$, where $x$ is the distance between $A$ and $B$ measured relative to $K$. ($A$ and $B$ are supposed of course to be fixed locations relative to $K$.) But the lower limit of topological simultaneity relative to $e_3$ is the event $\bar{e}_1$ which precedes $e_4$ by a time $x/c$, measured with respect to $K$. Therefore the condition for metrical simultaneity relative to $K'$ not to violate topological simultaneity is that

$$vx/c^2 < x/c$$

or

$$v < c,$$

but $v$ always *is* less than $c$ (since we are not envisaging the possibility of superluminal frame velocities), so the theorem is proved. Now the IP depends on adopting the Einstein convention relative to both $K'$ and $K$, so it is licenced by the above theorem, which in turn depends on the causal analysis of the simultaneity relation.

It is very instructive for our subsequent discussion to see what would happen if we adopted a synchronization convention in $K'$ that made a velocity other than that of light have equal values in oppositely opposed directions. For example we might consider 'acoustic relativity' as opposed to 'optical relativity'. Suppose that relative to $K$ a sound wave propagates with velocity $\pm w$ along the positive and negative $X$-axis respectively. Relative to $K'$ let us choose our synchronization convention in such a way that sound still appears to propagate 'isotropically'. It is clear that the appropriate choice for $m$ is just

$$(8) \qquad m = v/w^2$$

instead of the 'optical relativity' result given by (7).[43] Substituting in (5) and using (6) we obtain the transformation equations for 'acoustic relativity'[44]

$$(9) \qquad x' = 1/\sqrt{1-v^2/c^2} \cdot (x - vt)$$

$$t' = \frac{\sqrt{1-v^2/c^2}}{1-v^2/w^2} \cdot (t - vx/w^2).$$

THEOREM. For 'acoustic relativity' the synchronization in $K'$ becomes bizarre for $v > (w/c) \cdot w$.

*Proof*. The condition for bizarre synchronization is clearly

$$vx/w^2 > x/c, \qquad \text{i.e.,} \qquad v > (w/c) \cdot w.$$

Suppose we had a situation in which sound propagated faster than light relative to the frame $K$. To fix our ideas suppose the equation for the pressure $\phi$ in the sound wave is as of the form

$$(10) \qquad \frac{1}{w^2} \frac{\partial^2 \phi}{\partial t^2} - \frac{\partial^2 \phi}{\partial x^2} = 0, \quad \text{with } w > c.$$

Then we could use this superluminal sound to identify a latitude in topological simultaneity at a distance $x$ of $\pm x/w$ about the event distinguished as metrically simultaneous at this location by invoking the acoustic synchronization convention of 'isotropic' sound propagation in both directions along the $X$-axis. But now suppose we introduce the Einstein 'optical' synchronization convention in such a universe. This latter will now be bizarre in the ontological sense specified by our superluminal acoustic signalling when

$$vx/c^2 > x/w$$

i.e., when

(11) $v > (c/w) \cdot c.$

Since $w > c$, the Einstein optical convention, will become bizarre for $v < c$ and *in this sense* the IP, based as it is on the Einstein 'optical' convention, cannot be sustained. Hence $w > c$ is inconsistent with SR, identified with IP.

Indeed we can still employ Equation (9) for $w > c$, and as $w \to \infty$ we obtain the following form of the transformation equations

$$(12) \quad x' = 1/\sqrt{1 - v^2/c^2} \cdot (x - vt)$$
$$t' = \sqrt{1 - v^2/c^2} \cdot t.$$

These are just the so-called Sjödin-Tangherlini form of the Lorentz transformations[45]. If infinitely fast sound waves were available to us for signalling these would be the uniquely prescribed form of the Lorentz transformation, and the Einstein 'optical' convention would now be bizarre for *all* frame velocities of $K'$ greater than zero, according to (11).

All this is perfectly correct, but Grünbaum, Sjödin and Salmon[46], to take three influential examples, all effectively assume that this analysis of the superluminal acoustic case[47] can be applied directly to rule out the possibility of tachyons. But this is quite wrong. Let me explain. (10) is not itself a Lorentz-invariant equation. We have already violated SR by writing down such an equation. So it is not surprising that physical effects based on such an equation violate SR! Of course the tachyons discussed by physicists arise from consideration of a *Lorentz-invariant* equation, typically the Klein–Gordon equation with an imaginary mass–parameter, viz.

$$(13) \quad \frac{1}{c^2}\frac{\partial^2 \phi}{\partial t^2} - \frac{\partial^2 \phi}{\partial x^2} - \mu^2 \phi = 0.$$

The solutions of this equation have quite different properties from the non-invariant Equation (10). When subjected to quantization (13) describes tachyons moving with *any* velocity between $c$ and $\infty$[48]. This is quite different from an equation such as (10) which described the propagation of sound waves with a *unique* velocity $w > c$. Suppose an observer stationary in the frame of reference $K$ attempts to establish 'absolute' synchronization, using infinite-velocity tachyons (these are usually referred to in the tachyon literature as transcendent tachyons and we shall adopt this locution). We suppose that these transcendent tachyons are identified as transcendent relative to the adoption of the Einstein convention ($\epsilon = \frac{1}{2}$) in the $K$ frame. In Figure 8 we illustrate to-and-fro signalling between locations $A$ and $B$ using 'nearly transcendent' tachyons, where the event $e_4$ at $A$, which is metrically simultaneous with $e_3$ at $B$ according to the Einstein convention, is 'squeezed' between the emission ($e_1$) and return ($e_2$) of the tachyon signal. (It is clearer for our purposes to keep the outgoing and returning tachyon with distinct but very close world-lines.)

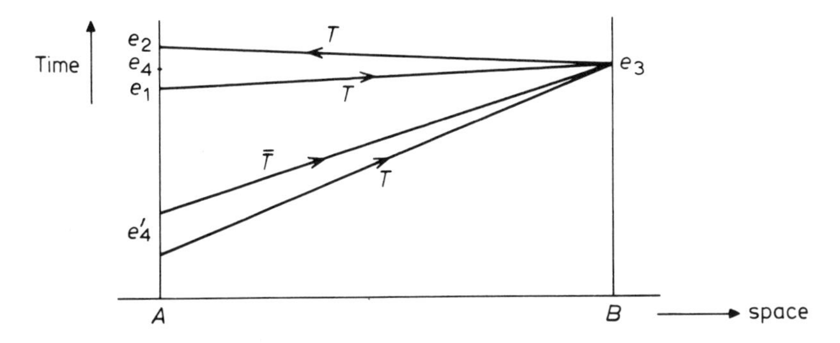


Fig. 8. Tachyon synchronization as viewed from $K$ ($T$ = tachyon, $\bar{T}$ = antitachyon).

But now consider the description of what is going on from the point of view of an observer attached to a frame $K'$, moving with uniform velocity along the positive $X$-axis, and for which we again assume the adoption of the Einstein synchronization convention (see Figure 9). He will deny that the physical processes being described relative to $K$ establish simultaneity between $e_3$ and $e_4$ at all. What is happening as seen from $K'$ is that a tachyon-antitachyon pair is emitted at $e_3$, the tachyon arriving at $e_2$ and the antitachyon at $e_1$.

The reason why, from the point of view of $K'$, no simultaneity relation is being established between $e_3$ and $e_4$ is that the chain of events $e_1 e_3 e_2$ is not a genidentical set, since two particles, not one, are clearly involved. Of course, relative to $K'$ there is another physical process which he, the observer attached to $K'$, describes as identifying $e_3$ as simultaneous with an event $e_4'$, 'squeezed' between tachyons travelling from $A$ to $B$ and back again with nearly transcendent velocities as viewed from $K'$. But the observer attached to $K$ will resist the claim that $e_3$ and $e_4'$ are truly simultaneous by pointing out that what he sees going on is a tachyon-antitachyon pair annihilating at $e_3$, as illustrated in Figure 8. Notice the symmetry in the situation

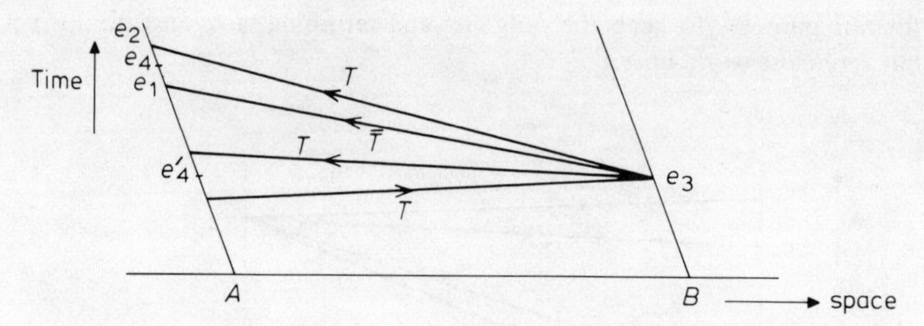


Fig. 9. Tachyon synchronization as viewed from $K'$.

– the observer in $K$ cannot *disallow* the description of his simultaneity claim by the observer in $K'$, since he is also going to invoke the Reinterpretation Principle to circumvent the $K'$ observer's claim that $e_3$ and $e_4{}'$ are simultaneous! In brief both observers must allow that what counts as a genidentical path from $e_1$ to $e_3$ and back to $e_2$ is a notion that depends on the observer and has no absolute significance[49]. In other words, it is only with bradyon or photon signalling that an unambiguous ascription of genidentity for the to-and-fro signal is available.

The difficulty with the Grünbaum notion of simultaneity, the relation we characterized as $S_G$, is now apparent. Reverting to the discussion given on p. 171, the definition of the betweenness relation $B$ and the simultaneity relation $S_G$ no longer satisfy the Meshing Condition. More specifically, it is the second part of the Meshing Condition that fails when tachyon interactions are taken into account. Now in a fundamental sense it is the Meshing Condition that we want to hold on to in capturing the notion of simultaneity. We must then replace $S_G$ with a 'revised' simultaneity relation $S_R$ that meets the Meshing Condition. A possible way of defining $S_R$ is as follows:

Let $e_5$ be a variable ranging over events at location $A$.

Then $e_3 S_R e_4$ iff $\forall e_5 \forall G(\sim G(e_3, e_4, e_5))$.

Clearly $e_3 S_G e_4 \rightarrow e_3 S_R e_4$ but not vice versa. $S_R$ will now satisfy the Meshing Condition provided we allow the general validity of

$$\exists e_5 \exists G(G(e_1, e_3, e_5)) \wedge \exists e_5 \exists G(G(e_2, e_3, e_5)) \\ \rightarrow \exists G(G(e_1, e_2, e_3)).$$

This condition is perfectly acceptable for interactions mediated by bradyons, photons *or* tachyons.

With the revised definition the range of events at location $A$, which are simultaneous with event $e_3$ at location $B$, is exactly the same as with photon signalling, and this is just the right latitude, as we have seen, to justify us in employing the Einstein convention in any (subluminal) frame $K'$ and hence underpins, ontologically speaking, the Invariance Principle. But we are still able to allow two events at space-like separation to be causally connectible, via a tachyon signal. We can literaly have our cake and eat it, if we are careful to distinguish causal connectibility from the violation of a simultaneity relation in the way we have sketched.

Let me stress again the key distinction between tachyons and our hypothetical superluminal phonons (acoustic particles). With infinite velocity phonons we *can* establish absolute synchrony. It is true of course that relative to some other frame $K'$ (which employs the Einstein optical synchronization convention) these phonons will no longer *appear* to have infinite velocity, but because the phonon equation *is not Lorentz-invariant* [50], these non-transcendent phonons cannot *also* occur as possible physical processes relative to the $K$-frame. It was this latter possibility that gave rise to the symmetrical situation as between $K$ and $K'$ in the tachyon case when disagreement arose as to whether $e_3$ was simultaneous with $e_4$ or $e_4'$. To put it another way, transcendent phonons pick out uniquely an absolute frame of reference relative to which they all have the same (infinite) velocity in all spatial directions. Tachyons, because they satisfy Lorentz-invariant equations, fail to pick out a preferred reference frame [51].

## 9. CONCLUSION

The immediate tension between SR and superluminal interaction arises on the Grünbaum–Reichenbach account from their version of the causal theory of time. In the preceding section we have endeavoured to remove this difficulty by revising the causal explication of the simultaneity relation. An alternative line of argument is of course to reject the whole apparatus of a causal theory of time [52]. In particular it is not clear that the notion of a continuous genidentical set of events can reasonably be taken as a primitive one. It is obviously not an observational primitive, and the notion of continuity may be regarded as grounded more naturally in a prior understanding of spatio-temporal order relations, in particular in the notion of spatio-temporal continuity, rather than the other way around. Consider the paradigm case of a ball rolling from $P$ to $P'$ and back again along the space-time track $a\,b\,e\,c\,d$ as sketched in Figure 10. ($P''$ is some arbitrary point between $P$ and $P'$.)

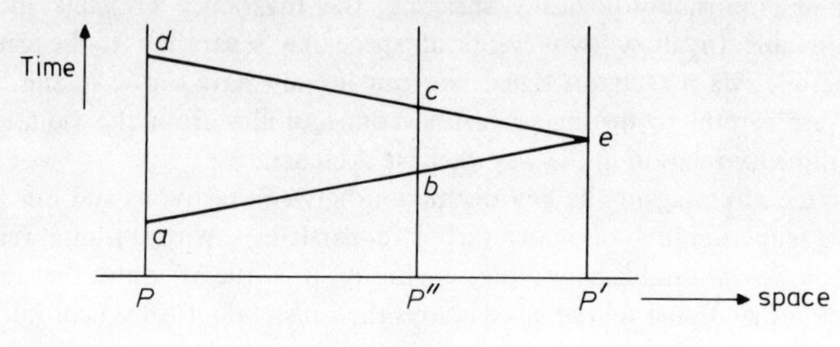


Fig. 10.

Considerations of purely *spatial* continuity would clearly not explain why the events between $a$ and $b$ together with those between $c$ and $d$ cannot be held to constitute a $k$-connection between $a$ and $d$. The reason for this is due to the *temporal* discontinuity which arises when we project this set onto the time axis (contrasting with the continuous projection onto the space axis). But in the causal theory of time we have to explain temporal discontinuity (an ordinal notion) in terms of the antecedently understood notion of a continuous $k$-connection. It may also be argued that the process of reidentifying a material object involves as a necessary (although not sufficient) condition an appeal to spatio-temporal continuity of its world-line.

At a more fundamental level the whole concept of a continuous world-line for a material particle becomes highly problematic at the level of quantal as opposed to classical description. In these circumstances we may want to look at SR as founded *directly* on the Invariance Principle in the sense that there exists a method of defining clock settings at any spatial location relative to any inertial frame such that the form of all physical laws remains invariant under the resulting transformation between relatively moving frames such as $K$ and $K'$[53]. But if this is all there is to SR, then the *possibility* of superluminal interaction must, as we have seen, be admitted.

In the context of reconciling the nonlocality we learned about from the EPR paradox with SR, the fact that we found no violation of Locality$_5$, that the Bell telephone was not possible, may be thought to tie in significantly with the limitations on tachyon-signalling capabilities inherent in our discussion of how to block the *additional* objection to superluminal interaction arising from causal loop paradoxes. In this paper I am not proposing any detailed mechanism whereby tachyon field theory could be used to 'explain' EPR nonlocality. A first attempt in that direction has been made by Fox (1972)[54]. My main objective has been to explore the locus of 'peaceful coexistence' between nonlocal features of QM and the constraints of SR at the level of *fundamental* objections to such a possibility. My conclusion is that many of the arguments dealing with these questions in the literature have overlooked essential features of the situation. The definitive resolution of this issue is one of the most challenging problems in current philosophy of physics.

## NOTES

[1] For further discussion of this point see Redhead (1981).

[2] Although $\Psi_{\text{singlet}}$ is a non-relativistic wave function this does *not* mean we cannot use arguments based on $\Psi_{\text{singlet}}$ to discuss conflict with SR. This is because $\Psi_{\text{singlet}}$ is the correct limiting form of a *relativistic* version of the two-body spin wave function relative to an appropriate frame of reference (centre-of-mass) where the two-body system is chosen to be one in which motion relative to the centre-of-mass is slow (e.g., low-energy proton-proton scattering).

[3] Of course there are observables such as $(\mathbf{S}(1) + \mathbf{S}(2))^2$ for which $\Psi_{\text{singlet}}$ is an eigenstate. When we are using measurements of $S_Z(1)$ *alone* to demonstrate the incompleteness of QM, this depends on the fact that $S_Z(2)$ does not commute with $(\mathbf{S}(1) + \mathbf{S}(2))^2$. So the interpretation of noncommutativity of the operators associated with observables in terms of the impossibility of ascribing simultaneous values to such observables, is still involved in the more economical version of the EPR argument referred to.

[4] We shall often loosely comflate an observable and its associated self-adjoint operator defined on the Hilbert space appropriate to the QM description of the system in question.

[5] We shall assume for the purposes of this paper that all observables have eigenvalues and associated eigenstates. Thus we shall gloss over the mathematical intricacies of dealing with operators with a continuous spectrum.

[6] The question of whether genuine comeasurability of noncommuting observables is possible (as distinct from state-preparations which simultaneously eliminate dispersion in the predicted measurement results of two noncommuting observables with no common eigenvectors) has been much discussed in the literature. See, for example, Popper (1934) and (1967), Arthurs and Kelly (1965), She and Heffner (1966), Park and Margenau (1968), Fitchard (1979). Proposed schemes involve measuring quantities with the same probability distributions as the observables in question, but it does not follow of course that, on a particular occasion, these latter observables have genuinely been measured. Also the resulting joint probability distributions which arise contradict the no-go theorems

of Cohen (1966) and Nelson (1967) which in turn stem from the 'Moyal problems' associated with attempts at defining joint probability distributions for noncommuting observables (see Moyal, 1949).

[7] We are here taking the line that propensities are properties of quantum-mechanical systems or objects, which are manifested or displayed in the context of repeatable arrangements or set-ups. Popper who introduced the propensity interpretation of probability to deal with the situation met with in QM (see, for example, Popper, 1959 and 1967) is not sympathetic to view (*B*) which he sees as arising from a mistaken attribution of propensities to objects rather than experimental arrangements. Of course, if propensities are introduced in the context of (*A*), such strictures seem to be in order, but if for other reasons (*B*) is more attractive then the manner of talking about propensities as properties of objects seems more natural and should lead to no confusion if the way such propensities are manifested is carefully spelled out in terms of repeatable set-ups. For further discussion of this point see Mellor (1971), Chapter 4, and Mackie (1973), pp. 179–187.

[8] The idea of potentiality is introduced, for example, by Heisenberg (1958), p. 53. These remarks are not really consistent with his general support of Bohr's Copenhagen interpretation of QM.

[9] See Margenau (1954) and McKnight (1958).

[10] We shall not in this work consider the so-called many-worlds interpretation of QM in which all potential results are actualized simultaneously in 'parallel' universes.

[11] See in particular the famous review article by Bell (1966). For a discussion of how to deal with the Kochen–Specker (1967) no-hidden-variables proof without violating realism, see van Fraassen (1973). A general discussion of the assumptions that lie behind the Kochen–Specker work and an evaluation of its critique by Fine (1973, 1974, 1977) has been given by the present author (Redhead, 1981).

[12] See in particular Wigner (1970), Bell (1971), Clauser and Horne (1974). In the last two references the possibility of stochastic hidden-variable theories in which measurement outcomes are not deterministically, but only stochastically, related to the specification of the hidden parameters, are also considered. We shall not specifically discuss such theories in the present work.

[13] Eberhard's proof has been improved technically by Peres (1978) and Brody (1980).

[14] As a typical example consider the case of correlating the signs of the spin-projections of two classical tops flying apart with zero total angular momentum and with their common spin axes isotropically distributed in space. It is easily shown that the correlation function is now given by $c(a, b) = -1 + 2\theta_{ab}/\pi$. See Bell (1964), Peres (1978).

[15] The correlation function referred to in Note 14 saturates the Bell inequality for all values of $\phi$. This is due to the special choice of an isotropic spatial distribution for the spin axes.

[16] In the case of stochastic hidden-variable theories referred to in Note 12, no specific assumption of joint distributions for noncommuting observables is made. See, however, the discussion in Fine (1980).

[17] A quite different approach to demonstrating violation of Locality$_3$ in QM is to consider two correlated systems with individual Hilbert spaces of dimension greater than two, for example two correlated spin-1 systems, and to show how the assumption of Locality$_3$ together with the application of Fine's Extended Value Rule (Fine, 1974, p. 284) restricted to comeasurable observables can be used to demonstrate a Kochen–

Specker paradox for value assignments to observables on one of the component systems. Similarly the inconsistency result proved by Fine in the appendix to his (1977) contains a hidden assumption of $\text{Locality}_3$ and (unsuspected by Fine) can actually be used to demonstrate violation of $\text{Locality}_3$. These ideas originate with the work of Simon Kochen (unpublished) and Stairs (1978). Further comment is given in Redhead (1981), Note 13, and Heywood and Redhead (1982).

[18] For detailed discussion and evaluation of the experimental tests of the Bell inequality see Clauser and Shimony (1978). These have mainly involved polarization correlations of photons, although there is one experiment by Lamehi–Rachti and Mittig (1976) involving spin-$\frac{1}{2}$ correlations as discribed in the text.

[19] See Aspect (1976). Even in this case bizarre possibilities of avoiding violations of Einstein locality remain – see the discussion in Clauser and Shimony (1978).

[20] The probability of decay at an instant of time is strictly zero. We should really speak of the probability of decay in a short interval $t_2 - \Delta$ to $t_2 + \Delta$ centred about $t_2$.

[21] This is just the usual Lewis analysis where I have ignored the case where the counterfactual is vacuously true in the sense that $\forall\, W_j(\sim W_j(\phi))$ holds. See Lewis (1973), p. 16.

[22] For further relevant discussion of tensed counterfactuals see Slote (1978), Lewis (1979) and Bowie (1979).

[23] I am using the terminology that a proposition is determinate if it is necessarily either true or false, but it may be undetermined in the sense that it is not either necessarily true or necessarily false. In the philosophical literature the locution 'determinate' is often used in the sense that I use 'determined'.

[24] See d'Espagnat (1976), pp. 58–61.

[25] Of course $\Psi_i$ and $\Psi_f$ differ in the predictions they make for the statistics of *correlation* measurements between particles 1 and 2, since with respect to such measurements $\Psi_i$ is not, while $\Psi_f$ is, an improper mixture.

[26] For a recent demonstration of this point see Ghirardi *et al.* (1980), who examine the flaws in proposed versions of what I have called the 'Bell' telephone. Further discussion of misunderstandings in the literature related to violations of $\text{Locality}_5$ in EPR experiments is given by Brown and Redhead (1981).

[27] This argument is *not* due to Shimony. It arose in discussion with my students.

[28] See for example Reichenbach (1957), pp. 129 and 205, and Grünbaum (1973), p. 827.

[29] For an early and influential discussion see Bilaniuk *et al.* (1962). Detailed development of the ideas was given in Feinberg (1967). For comprehensive references to the literature on tachyons see Recami (1978) and Caldirola and Recami (1981). A rather different approach was initiated by Bludman and Ruderman (1968) who considered the possibility of superluminal propagation of sound in ultra-dense matter. For further discussion of this work see Fox *et al.* (1970).

[30] We do not always distinguish a velocity (a vector) from its magnitude. The meaning should be clear from the context.

[31] Einstein (1905), pp. 63–64 in the Perrett and Jeffery translation.

[32] *Ibid*., p. 48.

[33] We assume $v < c$, i.e., the reference frames are composed of bradyons. We shall not discuss so-called 'extended relativity' involving *superluminal frames*.

[34] This is a converse of the familiar Feynman–Wheeler interpretation of a positron (antielectron) as a negative-energy electron propagating backwards in time.

35 We are here invoking the equivalence of the active and passive interpretation of the Lorentz transformation. The Lorentz-transform of a physically possible process relative to a *fixed* frame is another possible process relative to the *same* frame.

36 For further discussion of how causal loop paradoxes may be frustrated in terms of limitations on the requisite tachyon signalling capability see in particular Bilaniuk and Sudarshan (1969), Csonka (1970), Rolnik (1972), and in a more critical vein Maund (1979).

37 See Grünbaum (1973), pp. 193–194. In the following I confine the discussion to what Grünbaum calls o-betweenness. Note that Grünbuam also allows a notion of quasi-genidentity for light rays (photons).

38 Malement (1977) has demonstrated an alternative version of the causal theory of time which picks out the value $\epsilon = \frac{1}{2}$. Grünbaum in his (1973) overlooked this possibility, but the question of whether Grünbaum's version of the simultaneity relation, $S_G$, or Malement's version is the more appropriate in the context of understanding the significance of SR, remains. Malement rules out $S_G$ by requiring that simultaneity be an equivalence relation. But one way of expressing the insight of SR is just in showing that simultaneity is *not* an equivalence relation. The causal definability of $\epsilon = \frac{1}{2}$ synchronization may be a necessary, but is by no means a sufficient, condition that this value obtains as a matter of fact rather than convention. In this paper we shall confine our discussion to the Grünbaum analysis.

39 See in particular Poldlaha and Navrátil (1966) for a discussion of why the relationship should be a linear one.

40 My analysis follows, in a simplified form, the ideas developed in Podlaha and Navrátil (1967) and Sjödin (1979) I am grateful to Jon Dorling for drawing my attention to Sjödin's work, which generalizes the results I had obtained myself.

41 For example for two events labelled 1 and 2 on the track of the clock

$$\begin{aligned} t_2' - t_1' &= C(t_2 - t_1) - m(x_2 - x_1) \\ &= C(t_2 - t_1)(1 - mv) \quad \text{since } x_1 = vt_1 \\ &\qquad\qquad\qquad\qquad\quad \text{and } x_2 = vt_2 . \end{aligned}$$

Whence

$$G = \frac{t_2 - t_1}{t_2' - t_1'} = \frac{1}{C(1 - mv)} .$$

So

$$C = \frac{1}{G(1 - mv)} ,$$

as in the text. Similar for A.

42 Thus from (4) the velocity transformation from $U_x$ to $U_x'$, say, is given by

$$U_x' = \frac{A}{C} \quad \frac{U_x - v}{1 - mU_x} .$$

Taking $U_x = \pm c$ and equating the resulting magnitudes for $U_x'$, yields at once the result given in the text.

[43] The derivation given in Note 42 goes through with $c$ replaced by $w$. The value of the average 2-way velocity of sound relative to $K'$ is of course no longer $w$, but is now given by

$$w' = \frac{1 - v^2/w^2}{1 - v^2/c^2} \cdot w.$$

The general relation between $m$ and the equivalent (acoustic) Reichenbach $\epsilon$-parameter is easily seen to be

$$m = \frac{v - (2\epsilon - 1)w}{w^2 - (2\epsilon - 1)wv}.$$

[44] As $c \longrightarrow \infty$, we recover the Zahar transformation (with $c$ replaced by $w$). Cf. Zahar (1977) who discusses the use of the Einstein convention in a Newtonian world.
[45] Sjödin (1979), Tangherlini (1961).
[46] For Salmon's discussion on tachyons see his (1975), pp. 122–124.
[47] I have of course filled in the details of the argument. The mathematics of acoustic relativity is not explicitly given by any of the authors cited, although it is implicit in the generalized analysis of Sjödin (1979).

The idea of acoustic relativity is sometimes referred to in the literature (see in particular Kar, 1970), but in a different sense from the one I am using, viz. acoustic synchronization in an Einsteinean universe.
[48] For the quantum field theory of tachyons see Feinberg (1967). For the technical difficulties encountered (not all of which were appreciated in the pioneering work of Feinberg) see Kamoi and Kamefuchi (1978). These stem essentially from the impossibility of defining a Lorentz-invariant vacuum state (since the distinction between absorbtion and emission of tachyons is no longer an absolute one but depends on the frame of reference as we explained in Section 7).
[49] The fact that to-and-fro signalling with tachyons is in conflict with the notion of genidentity can be more sharply expressed in two-dimensional space-time. If we take as necessary condition for the identity of a particle that its successive states be connectible by an element of the proper Poincaré group, then in one-dimensional space a tachyon cannot 'turn round', i.e., there is an invariant distinction between tachyons moving to the left say and tachyons moving to the right. With three-dimensional space, however, a tachyon can make a U-turn without a change of identity. See for example Antippa and Everett (1971).
[50] To avoid any possibility of misunderstanding I am not of course suggesting that one cannot give a relativistic theory of sound waves. But to do this we have to take account of the possible motions of the medium. Trivially a velocity boost generates a possible motion, viz., one in which a 'wind' is blowing!
[51] I here disagree with Grünbaum's interpretation of Bilaniuk and Sudarshan's discussion of a preferred reference frame as being required *physically* in tachyon field theory. See Grünbaum (1973), p. 827, and Bilaniuk and Sudarshan (1969).
[52] For a critical discussion of the causal theory of time see for example Lacey (1968).
[53] If this principle is true we may infer that there is no *factual* way of distinguishing $K$ and $K'$, and noticing that the clock-synchronization procedures which make IP possible

lead directly to the relativity of simultaneity, we arrive at the idea that absolute time must be replaced by the notion of time-relative-to-a-frame. Notice that IP does not necessitate choosing the Einstein convention ($\epsilon = \frac{1}{2}$) in both frames, only that the same $\epsilon$-value is used in $K$ and $K'$. Of course with $\epsilon \neq \frac{1}{2}$ the form of the physical laws becomes more complex, for example electromagnetic waves no longer propagate isotropically in free space, but these more complex laws will still exhibit form-invariance under the appropriate $\epsilon$-extension of the Lorentz transformations.

[54] I also do not want to gloss over the technical problems associated with tachyon field theory, in particular the difficulty of reconciling the RIP with unitarity, which are well set out in Kamoi and Kamefuchi (1978), cited in Note 48. However, these may reflect on the machinery of quantum field theory rather than on the possible existence of tachyons. In passing I may add that preliminary reports of observing tachyons, as in Clay and Crouch (1974), have not been substantiated by later experimental work.

## REFERENCES


Antippa, A. F. and Everett, A. E.: 1971, 'Tachyons without Causal Loops in One Dimension', *Physical Review* **D4**, 2198–2203.

Arthurs, E. and Kelly, J. L.: 1965, 'On the Simultaneous Measurement of a Pair of Conjugate Observables', *Bell System Technical Journal* **44**, 725–729.

Aspect, A.: 1976, 'Proposed Experiment to Test Nonseparability of Quantum Mechanics', *Physical Review* **D4**, 1944–1951.

Bell, J. S.: 1964, 'On the Einstein–Podolsky–Rosen Paradox', *Physics* **1**, 195–200.

Bell, J. S.: 1966, 'On the Problem of Hidden Variables in Quantum Mechanics', *Reviews of Modern Physics* **38**, 447–452.

Bell, J. S.: 1971, 'Introduction to the Hidden-Variable Question', in B. d'Espagnat (ed.), *Foundations of Quantum Mechanics* (Proceedings of the International School of Physics 'Enrico Fermi', Course IL, Academic, New York), pp. 171–181.

Benford, G. A., Book, D. L. and Newcomb, W. A.: 1970, 'The Tachyonic Antitelephone', *Physical Review* **D2**, 263–265.

Bilaniuk, O. M. P., Deshpande, V. K., and Sudarshan, E. C. G.: 1962, '"Meta" Relativity', *American Journal of physics* **30**, 718–723.

Bilaniuk, O. M. P. and Sudarshan, E. C. G.: 1969, 'More about Tachyons – The Rebuttal', *Physics Today* **22**, 30–52.

Bludman, S. A. and Ruderman, H. A.: 1968, 'Possibility of the Speed of Sound Exceeding the Speed of Light in Ultradense Matter', *Physical Review* **170**, 1176–1184.

Bohm, D.: 1951, *Quantum Theory*, Prentice-Hall, Englewood Cliff, N. J.

Bohr, N.: 1935, 'Can Quantum-Mechanical Description of Physical Reality be Considered Complete?', *Physical Review* **48**, 696–702.

Bowie, G. L.: 1979, 'The Similarity Approach to Counterfactuals: Some Problems', *Noûs* **13**, 477–498.

Brody, T. A.: 1980, 'Where Does the Bell Inequality Lead?', unpublished manuscript.

Brody, T. A. and De La Peña-Auerbach, L.: 1979, 'Real and Imagined Nonlocalities in Quantum Mechanics', *Il Nuovo Cimento* **54B**, 455–462.

Brown, H. R. and Redhead, M. L. G.: 1981, 'A Critique of the Disturbance Theory of Indeterminacy in Quantum Mechanics', *Foundations of Physics* **11**, 1–20.

Caldirola, P. and Recami, E.: 1981, 'Causality and Tachyons in Relativity', in M. L. Dalla Chiara (ed.), *Italian Studies in the Philosophy of Science*, D. Reidel, Dordrecht, pp. 249–298.

Clauser, J. F. and Horne, M. A.: 1974, 'Experimental Consequences of Objective Local Theories', *Physical Review* **D10**, 526–535.

Clauser, J. F. and Shimony, A.: 1978, 'Bell's Theorem: Experimental Tests and Implications', *Reports on Progress in Physics* **41**, 1881–1927.

Clay, R. W. and Crouch, P. C.: 1974, 'Possible Observation of Tachyons Associated with Extensive Air Showers', *Nature* **248**, 28–30.

Cohen, L.: 1966, 'Can Quantum Mechanics be Formulated as a Classical Probability Theory?', *Philosophy of Science* **33**, 317–322.

Csonka, P. L.: 1970, 'Causality and Faster than Light Particles', *Nuclear Physics* **B21**, 436–444.

D'Espagnat, B.: 1976, *Conceptual Foundations of Quantum Mechanics*, 2nd ed., Benjamin, Reading, Mass.

Eberhard, P. H.: 1977, 'Bell's Theorem without Hidden Variables', *Il Nuovo Cimento* **38B**, 75–80.

Einstein, A.: 1905, 'Zur Electrodynamik bewegten Körper', *Annalen der Physik* **17**, 891–921. English translation by W. Perrett and G. B. Jeffery, in A. Einstein *et al.*, *The Principle of Relativity*, Methuen, London, 1923. Reprint ed., Dover, New York, 1951.

Einstein, A., Podolsky, B. and Rosen, N.: 1935, 'Can Quantum-Mechanical Description of Physical Reality be Considered Complete?', *Physical Review* **47**, 777–780.

Feinberg, G.: 1967, 'Possibility of Faster-Than-Light Particles', *Physical Review* **159**, 1089–1105.

Fine, A.: 1973, 'Probability and the Interpretation of Quantum Mechanics', *The British Journal for the Philosophy of Science* **24**, 1–37.

Fine, A.: 1974, 'On the Completeness of Quantum Theory', *Synthese* **29**, 257–289. Reprinted in P. Suppes (ed.): *Logic and Probability in Quantum Mechanics*, D. Reidel, Dordrecht, 1976, pp. 249–281.

Fine, A.: 1977, 'Conservation, the Sum Rule and Confirmation', *Philosophy of Science* **44**, 95–106.

Fine, A.: 1979, 'Counting Frequencies: A Primer for Quantum Realists', *Synthese* **42**, 145–154.

Fine, A.: 1980, 'Correlations and Physical Locality', forthcoming in P. D. Asquith and R. N. Giere (eds.), *PSA: 1980, Vol. II* (Proceedings of the 1980 Biennal Meeting of the Philosophy of Science Association), Philosophy of Science Association, East Lansing.

Fitchard, E. E.: 1979, 'Proposed Experimental Test of Wave Packet Reduction and the Uncertainty Principle', *Foundations of Physics* **9**, 525–535.

Fox, R.: 1972, 'Tachyons and Quantum Statistics', *Physical Review* **D5**, 329–331.

Fox, R., Kuper, C. G., and Lipson, S. G.: 1970, 'Faster-than-Light Group Velocities and Causality Violation', *Proceedings of The Royal Society* (*London*), Series A, 316, pp. 515–523.

Ghirardi, G. C., Rimini, A., and Weber, T.: 1980, 'A General Argument against Superluminal Transmission through the Quantum Mechanical Measurement Process', *Lettere al Nuovo Cimento* **27**, 293–298.

Grünbaum, A.: 1973, *Philosophical Problems of Space and Time*, 2nd ed., D. Reidel, Dordrecht.

Heisenberg, W.: 1958, *Physics and Philosophy: The Revolution in Modern Science*, Harper, New York.

Heywood, P. and Redhead, M. L. G.: 1982, 'Nonlocality and the Kochen–Specker Paradox', forthcoming.

Kamoi, K. and Kamefuchi, S.: 1978, 'Tachyons as Viewed from Quantum Field Theory', in E. Recami (ed.), *Tachyons, Monopoles and Related Topics*, North-Holland, Amsterdam, pp. 159–167.

Kar, K. C.: 1970, 'Relativity in the Acoustical World', *Indian Journal of Theoretical Physics* **18**, 1–11.

Kochen, S. and Specker, E.: 1967, 'The Problem of Hidden Variables in Quantum Mechanics', *Journal of Mathematics and Mechanics* **17**, 59–87. Reprinted in C. A. Hooker (ed.), *The Logico-Algebraic Approach to Quantum Mechanics*, Vol. I: *Historical Evolution*, D. Reidel, Dordrecht, 1975, pp. 293–328.

Lacey, H. M.: 1968, 'The Causal Theory of Time: A Critique of Grünbaum's Version', *Philosophy of Science* **35**, 332–354.

Lamehi-Rachti, M. and Mittig, W.: 1976, 'Quantum Mechanics and Hidden Variables: A Test of Bell's Inequality by the Measurement of the Spin Correlation in Low-Energy Proton-Proton Scattering', *Physical Review* **D14**, 2543–2555.

Lewis, D.: 1973, *Counterfactuals*, Blackwell, Oxford.

Lewis, D.: 1979, 'Counterfactual Dependence and Time's Arrow', *Noûs* **13**, 455–476.

Mackie, J. L.: 1973, *Truth, Probability, and Paradox*, Oxford University Press, Oxford.

McKnight, J. L.: 1958, 'An Extended Latency Interpretation of Quantum Mechanical Measurement', *Philosophy of Science* **25**, 209–222.

Malament, D.: 1977, 'Causal Theories of Time and the Conventionality of Simultaneity', *Noûs* **11**, 293–300.

Margenau, H.: 1954, 'Advantages and Disadvantahes of Various Interpretations of the Quantum Theory', *Physics Today* **7**, 6–13.

Maund, J. B.: 1979, 'Tachyons and Causal Paradoxes', *Foundations of Physics* **9** 557–574.

Mellor, D. H.: 1971, *The Matter of Chance*, Cambridge University Press, Cambridge.

Moyal, J. E.: 1949, 'Quantum Mechanics as a Statistical Theory', *Proceedings of the Cambridge Philosophical Society* **45**, 99–124.

Nelson, E.: 1967, *Dynamical Theories of Brownian Motion*, Princeton University Press, Princeton, N. J.

Newton, R. G.: 1967, 'Causality Effects of Particles that Travel Faster than Light', *Physical Review* **162**, 1274.

Park, J. L. and Margenau, H.: 1968, 'Simultaneous Measurability in Quantum Theory', *International Journal of Theoretical Physics* **1**, 211–283.

Parmentola, J. A. and Yee, D. D. H.: 1971, 'Peculiar Properties of Tachyon Signals', *Physical Review* **D4** 1912–1915.

Peres, A.: 1978, 'Unperformed Experiments Have no Results', *American Journal of Physics* **46**, 745–747.

Pirani, F. A. E.: 1970, 'Noncausal Behaviour of Classical Tachyons', *Physical Review* **D1**, 3224–3225.

Podlaha, M. and Navrátil, E.: 1966, 'On the Linearity of the Lorentz Transformation', *Acta Physica Austriaca* **24**, 99–100.

Podlaha, M. and Navrátil, E.: 1967, 'Formulation of the Axioms for Deriving the Lorentz Transformations, Galileo Transformations and Palacios Transformations', *Revista de la Real Academia de Ciencias Exactas, Fisica y Naturalis de Madrid* **61**, 555–561.

Popper, K. R.: 1934, *Logik der Forschung*, Springer-Verlag, Vienna. English translation: *The Logic of Scientific Discovery*, Hutchinson, London, 1959.

Popper, K. R.: 1959, 'The Propensity Interpretation of Probability', *The British Journal for the Philosophy of Science* **10**, 24–42.

Popper, K. R.: 1967, 'Quantum Mechanics without "The Observer"', in M. Bunge (ed.), *Quantum Theory and Reality*, Springer-Verlag, Berlin-Heidelberg-New York, pp. 7–44.

Recami, E.: 1978, 'An Introductory View about Superluminal Frames and Tachyons', in E. Recami (ed.), *Tachyons, Monopoles and Related Topics*, North-Holland, Amsterdam, pp. 3–25.

Redhead, M. L. G.: 1981, 'Experimental Tests of the Sum Rule', *Philosophy of Science* **48**, 50–64.

Reichenbach, H.: 1957, *The Philosophy of Space and Time*, English translation by H. Reichenbach and J. Freund, Dover, New York.

Rolnick, W. B.: 1969, 'Implications of Causality for Faster-than-Light Matter', *Physical Review* **183**, 1105–1108.

Rolnick, W. B.: 1972, 'Tachyons and the Arrow of Causality', *Physical Review* **D6**, 2300–2301.

Salmon, W. C.: 1975, *Space, Time and Motion*, Dickenson, Encino, California.

She, C. Y. and Heffner, H.: 1966, 'Simultaneous Measurement of Noncommuting Observables', *Physical Review* **152**, 1103–1110.

Shimony, A.: 1978, 'Metaphysical Problems in the Foundations of Quantum Mechanics', *International Philosophical Quarterly* **18**, 3–17.

Sjödin, T.: 1979, 'Synchronization in Special Relativity and Related Theories', *Il Nuovo Cimento* **51B**, 229–246.

Slote, M. A.: 1978, 'Time in Counterfactuals', *The Philosophical Review* **87**, 3–27.

Stairs, A.: 1978, *Quantum Mechanics, Logic and Reality*, Unpublished Ph.D. Thesis, University of Western Ontario.

Stapp, H. P.: 1971, '*S*-Matrix Interpretation of Quantum Theory', *Physical Review* **D3**, 1303–1320.

Tangherlini, F. R.: 1961, 'An Introduction to the General Theory of Relativity', *Supplemento Nuovo Cimento* **20**, 1–86.

Tolman, R. C.: 1917, *The Theory of Relativity of Motion*, University of California Press, Berkeley.

Van Fraassen, B.: 1973, 'Semantic Analysis of Quantum Logic', in C. A. Hooker (ed.), *Contemporary Research in the Foundations and Philosophy of Quantum Theory*, D. Reidel, Dordrecht, pp. 80–113.

Wigner, E. P.: 1970, 'On Hidden Variables and Quantum Mechanical Probabilities', *American Journal Journal of Physics* **38**, 1005–1009.

Zahar, E.: 1977, 'Mach, Einstein and the Rise of Modern Science', *The British Journal for the Philosophy of Science* **28**, 195–213.

PETER GIBBINS

# QUANTUM LOGIC AND ENSEMBLES

Probability, as it appears in quantum mechanics, is a measure on the non-boolean lattice of propositions known as *quantum logic.* It is therefore no surprise that quantum probability is non-classical and that attempting to impose on quantum probability a classical ensemble interpretation leads to paradox. One such paradox is provided by the ignorance interpretation of mixtures, another by the Einstein–Podolsky–Rosen (EPR) thought-experiment. I shall argue, in reply to Dr. Redhead, that a *quantum logical ensemble* (QLE) interpretation of quantum mechanics resolves these paradoxes in a natural way and that in the case of the EPR paradox one need not invoke superluminal causal connections to account for the non-locality. This paper therefore has somewhat narrower scope than Dr. Redhead's for I shall have no need to examine tachyonic mechanisms to explain the EPR correlations.

## I. THE EINSTEIN–PODOLSKY–ROSEN PARADOX

A naive interpreter of quantum mechanics will find, among the applications of the theory, a good many instances of non-locality. Identifying the individual system with the square-modulus of its wave-function (in the style of Schrodinger), and accepting the reduction of the wave-packet during measurment, he finds that a successful position measurement localising the system *here* generates its total disappearance from *there* at the instant of measurement. Immediately after the measurement the system will re-appear (almost) *everywhere* since the position operators $\hat{\mathbf{r}}_t$ and $\hat{\mathbf{r}}_{t'}$ are totally incompatible when $t \neq t'$. There is, in other words, no state-vector that the commutator $[\hat{\mathbf{r}}_t, \hat{\mathbf{r}}_{t'}]$ sends into the zero-vector. This instantaneous disappearing and re-appearing involves action at a distance and seems to violate special relativity, as Einstein himself pointed out early on in his debate with Niels Bohr.

Alternatively, imagining as Born originally did, the individual system to be a particle with simultaneous precise values for all its dynamical variables, we again run into non-locality. An obvious example is provided by the two-slit experiment. How can the open slit that the particle does not go through affect the probability of its hitting the screen at the small region $y_R$? Only,



*R. Swinburne (ed.), Space, Time and Causality*, 191–205.

it would seem, by means of an action at a distance by the slit on the particle. The Pauli Principle provides another example. If electrons are localised particles what mechanism, other than action at a distance, prevents the two spatially separated electrons in a helium atom adopting the same state?

The lesson to be learnt from this is that naive interpretations of quantum mechanics lead to difficulties, of which non-locality is one. The great merit of the Einstein–Podolsky–Rosen paradox, or *thought-experiment*, (hereafter EPR) is that it exhibits a form of non-locality not dependent on naivety of interpretation.

In their original paper of 1935, Einstein, Podolsky and Rosen consider a two-particle system described in one dimension at a fixed moment in time by a wave-function $\psi(x_I, x_{II})$ – the $x_I$ and $x_{II}$ are the position variables for the two systems I and II respectively – the wave-function being represented as an integral over the product of the momentum 'eigenstates' of the two particles

$$\psi(x_I, x_{II}) \sim \int_{-\infty}^{\infty} \exp 2\pi_i(x_I - x_{II} + x_0)p/\hbar]\, dp.$$

A measurement of the momentum of particle I will result in some value $p$. This measurement on particle I will collapse $\psi(x_I, x_{II})$ into a product of two terms, the first a momentum eigenstate of particle I having the eignevalue $p$, and the second, a momentum eigenstate of particle II having the eigenvalue $-p$. Alternatively, a position measurement on particle I will yield some value $x$ for its position. This will force the position eigenstate corresponding to the position eigenvalue $(x + x_0)$ on to particle II.

These results have two disturbing features. First, a measurement on particle I can affect the eigenstate of particle II, from a distance. But secondly, and more importantly, measurements made on particle I can put particle II into *either* an eigenstate of momentum *or* an eigenstate of position, depending on the whim of the observer making measurements on particle I. The EPR conclusion is that *either* there is action at a distance of measurements on particle I on to particle II *or* that quantum states are incomplete, that is, that particle II carried information as to how it should behave under position and momentum measurements, information not revealed in the quantum mechanical formalism. EPR favoured the second of these disjuncts, and ended their paper with a commitment to a *hidden-variables* underpinning of quantum mechanical states.

One can make a number of technical objections to the EPR account. And

Bohm's reworking of it (Bohm, 1951, p. 614ff.), which we call EPR-B following Popper (1971), has become the standard replacement. Among the objections are that EPR uses the disreputable apparatus of position and momentum eigenstates, neither of which exist. If the two particles are both electrons, then $\psi(x_{\text{I}}, x_{\text{II}})$ is not appropriately anti-symmetrised. And one may criticise the behaviour of $\psi(x_{\text{I}}, x_{\text{II}})$ for its behaviour under the evolution operator (see Redhead, 1981, p. 58ff.).

But there *is* not respect in which EPR might appear to be superior to EPR-B. And that is that EPR deals explicitly with the wave-function for the two particles as represented in configuration-space. The separation of the two particles appears explcitly in the original EPR account. In EPR-B it is implicit. We *assume* that the spin measurements on I and II are spatially separated. One aspect of this can cause confusion. The EPR-B singlet state is represented by a state-vector in the tensor product of the Hilbert Spaces associated with the two particles. The spatial parts of these distinct state-vectors, one for each separate particle, will overlap. One might be tempted to think that this overlap enables some mechanism of interconnection to operate. However, the EPR-B example shows that the correlation between the two particles is quite independent of the magnitude of the overlap, precisely because in EPR-B the overlap is not mentioned in the example at all.

Let us consider the EPR-B account explicitly. Consider the decomposition of a diatomic molecule of total spin zero (i.e., in the singlet state) into two atoms each of spin $\frac{1}{2}\hbar$. The molecular example is, of course, equivalent to that of two interacting electrons. One represents the singlet state of the molecule by the wave-function $\psi$ and the spin states of the atoms I and II by the single particle state-vectors $|\pm, \delta, \text{I}\rangle$ and $|\pm, \delta, \text{II}\rangle$. So that

$$\psi(\text{I}, \text{II}) = (\tfrac{1}{2})^{1/2}\,[|+, \delta, \text{I}\rangle \otimes |-, \delta, \text{II}\rangle - |-, \delta, \text{I}\rangle \otimes |+, \delta, \text{II}\rangle]$$

for any direction $\delta$. Rotation of $\delta$ into a new direction $\delta'$ produces a new representation of $\psi$ having the same form, $\delta'$ replacing $\delta$.

The atoms I and II will separate after the decomposition. A $\pm$ result for a spin measurement on particle I yields the information that a similar measurement on particle II will certainly give a $\mp$ result in the same direction. But we are free to choose the direction for the spin measurement on I. No quantum mechanical state is a spin $\frac{1}{2}\hbar$ eigenstate for two distinct directions. It seems therefore that either system II is predetermined to respond to spin measurements in a way not captured by quantum mechanical ascriptions of state, or that there is some action at a distance from the measurement on system I to the system II.

Before we consider how this example is handled in a quantum logical ensemble interpretation it is worth considering our reason for thinking that particles I and II really are separate.

EPR sometimes encourages the metaphysical thought that the two particles I and II lose their separate identities, and that the 'paradox' is the result of our failing to recognise this fact. On such a view we have, after the decomposition, a 'two-atom situation' rather than two separate atoms. The EPR case will then look rather more like a case of the 'collapse of the wave-packet' non-locality discussed at the beginning of Section I.

There is, I shall argue, a sense in which elementary quantum mechanics, in its customary formulation, presupposes that the two particles retain their separate identities. And there is a sense in which the *second quantisation formalism* presupposes the 'two-atom situation' picture. EPR is, of course, formulated within elementary quantum mechanics, and the interpretation offered here is an interpretation of that more elementary theory.

First, elementary quantum mechanics. For fermions, half-integral spin particles, one asserts that

$$(*) \qquad \psi(\mathrm{I}, \mathrm{II}) = -\psi(\mathrm{II}, \mathrm{I}).$$

When one interchanges the particles the wave-function changes sign. This assertion would be unintelligible if the labels I and II were not to refer to particles which retain their identity. (I am not here claiming that indistinguishable particles may be separated in observation.)

But in the second quantisation formalism the role of this assertion (*) is taken by the anticommutation rules for the particle *creation* and *annihilation operators*. The state of the two-particle system is represented by a state-vector in *Fock Space* and the *number* of particles in the state is an observable, with a correspond-operator. There is no need to distinguish between the two particles. We however follow the account given by elementary quantum mechanics.

## II. DENSITY OPERATORS AND STATES

It is often said that quantum mechanics modifies the classical concepts of both *state* and *observable*. Taking an orthodox individual system view of quantum mechanics, one says that quantum mechanical states are represented by state-vectors in Hilbert Space, while the observables are represented by operators on that space.

But there is a more general notion of quantum mechanical state.

Speaking *subjectively*, suppose I try to assign a state-vector to some system. I assume that some state-vector or other does represent the state of the system but, because I am unsure what that state-vector is, I choose to represent the state of the system as a weighted sum over many distinct state-vectors $|j\rangle$. The weights $\lambda_j$ associated with each state-vector will then give my subjective probability that the system has that state-vector. The operator

$$\hat{\rho} = \sum_j \lambda_j |j\rangle\langle j|, \qquad \lambda_j \geqq 0, \quad \sum_j \lambda_j = 1.$$

is called the *density operator* for the system.

An *objective* interpretation of the density operator might run as follows. The density operator $\hat{\rho}$ might be held to describe a quantum ensemble which is decomposable into disjoint homogeneous subensembles (we here assume all the $|j\rangle$ are orthonormal), each having 'relative size' given by its weighting $\lambda_j$. One assumes that the density operator for each homogeneous or pure subensemble has just one term in the sum

$$\hat{\rho}_j = |j\rangle\langle j|.$$

An ensemble, or subensemble, represented by such a density operator is said to be a *pure case*. An inhomogeneous ensemble is said to be a *mixture*. There is a simple mathematical distinction between pure cases and mixtures. For a pure case $\hat{\rho}^2 = \hat{\rho}$. For a mixture $\hat{\rho}^2 < \hat{\rho}$.

It is very natural to associate quantum states with density operators. For then the state-vector description comes out as a special case. The quantum mechanical algorithm is expressible in a particularly simple way. For any observable $A$, the expectation value

$$\langle \hat{A} \rangle = \text{Trace}\,[\hat{\rho}\hat{A}]$$

The objective interpretation of the density operator associates it with an ensemble and associates the *eigenvalues* of the density operator – the $\lambda_j$'s – with the weightings of the corresponding pure subensembles in the ensemble. It is usual to take this interpretation one stage further and append the following thesis.

($I^2M$) Each individual system belongs to one and only one pure subensemble in a mixture described by the density operator $\hat{\rho}$. The eigenvalues of $\hat{\rho}$ then express a *measure of our knowledge* that a chosen individual system belongs to the corresponding pure subensemble.

This thesis, ($I^2M$), is naturally called *the ignorance interpretation of mixtures*.

All but the most fastidious (or 'minimal') ensemble interpretations make the following additional claim.

(IS) Whenever an individual system belongs to a pure (sub)ensemble it, the individual system, has those properties assigned by quantum theory to the ensemble.

These apparently innocuous principles lead to paradox.

Consider a density operator which is *degenerate*, that is, in which $\lambda_i = \lambda_j$ for some $i \neq j$. An extremely simple example is that of of the density operator $\hat{\rho}_0$ describing an ensemble of electrons whose spins are randomly orientated. Such a density operator may be taken to represent two equally weighted pure subensembles of electrons whose spins are aligned in the 'up' and 'down' senses respectively *in any chosen spatial direction* $\delta$.

(#) $$\hat{\rho}_0 = \tfrac{1}{2}\,|+, \delta><+, \delta\,| + \tfrac{1}{2}\,|-, \delta><-, \delta\,|$$

Whatever direction $\delta$ chosen for the eigenstates $|\stackrel{+}{-}, \delta>$, the density operator remains the same. Only its representation changes. The ignorance interpretation of mixtures generates the paradox that the ensemble described by $\hat{\rho}_0$ consists of two equally weighted pure subensembles of spin-up and spin-down respectively in (for example) the $z$-direction, and two similarly pure subensembles of spin-up and spin-down in (for example) the $x$-direction. But no individual system can belong to more than one of these four sub-ensembles. It is as if one were saying that an urn contained a collection of balls each of which is either black or white, *and* either red or green.

One might conclude that the density operator description of ensembles is fundamentally *incomplete*, that statistically equivalent descriptions (like all the non-denumerably infinite possible representations of $\hat{\rho}_0$) are not all equally good. This is one of the lines taken by Grossman (Grossman, 1974). For Grossman argues that only one of this non-denumerable infinity of orthogonal decompositions corresponds to the *actual* decomposition of the mixture into pure subensembles. This *privileged* decomposition is to be determined by that pragmatic feature of the ensemble which is the history of its preparation, rather than any formal property, such as its statistical behaviour.

Following van Fraassen (van Fraassen, 1972), and thereby Grossman, the arguments against the ignorance interpretation of mixtures [1]

may be roughly divided into (a) those emphasising that the ignorance interpretation is unrealistic with respect to physical situations and (b) those claiming that, if added to the basic principles of quantum theory, the ignorance interpretation leads to inconsistency.

Against arguments of class (a) Grossman makes two claims. First, consider the case where we have a density operator such as $\hat{\rho}_0$. Suppose such an ensemble was *prepared* by mixing together two pure ensembles of spin-up in the $z$-direction and spin-down in the $z$-direction. We may represent $\rho_0$ using the basis $|\pm, \delta>$ for any chosen direction $\delta$. But according to Grossman, there is a natural choice of $\delta$, namely $\delta = z$. This is the natural choice for the following *physical* reason. In any physically realistic preparation procedure, we cannot be sure that the two ensembles are going to be mixed with exactly equal weightings. And, Grossman argues, it is only for exactly equally weighted subensembles in the final mixture, that is only for *degenerate* $\hat{\rho}_0$, that the orthogonal decomposition will fail to be unique. We should, according to Grossman, represent the prepared mixture by the density operator $\rho_0'$ where

$$(\#\#) \qquad \hat{\rho}_0' = (\tfrac{1}{2} + \epsilon)\,|+, z><+, z\,| + (\tfrac{1}{2} - \epsilon)\,|-, z><-, z\,|$$

where $|\epsilon| \ll 1$. Any 'real collection of electrons' must therefore be described by an expansion of the form (##).

In other words, because we are ignorant of the exact value of $\hat{\rho}_0$ we should choose a particular representation of $\hat{\rho}_0$ such that the form of this representation is not sensitive to small changes in the mixture. That is, we should choose that representation of $\hat{\rho}_0$ such that small changes in the mixture do not introduce *cross-terms* in the expansion.

There are, however, two points that should be made against this pragmatic evasion of the paradox. The first is that we are concerned with a problem of interpretation. *Given* that we have a density operator of the form (#), may be sustain the ignorance interpretation? It is not a sufficient response to claim that such a density operator can be held not to arise *in practice*. Secondly, there is the problem of *ensembles* and *beams*.[2] By a beam, we mean a collection of real individual systems. By an ensemble we mean a non-denumerable collection of objects. Only if the cardinality of an ensemble is non-denumerable can the axioms of probability be strictly true of it. So that a beam cannot be an ensemble. An ensemble is an abstract (or 'virtual' in the terminology of Gibbsian statistical mechanics) object. We cannot therefore speak of *preparing* an ensemble. We can of course prepare a beam. The behaviour of a beam approximates to that of an ensemble in the sense that averages for the beam approximate to expectation-values for the ensemble. Pragmatic considerations arise not when we consider the representation for an ensemble, but rather when we consider which ensemble we should use to approximate to the beam. Quantum mechanics describes ensembles. It is no way out of a problem of interpretation to say that the problematic objects may be avoided in practice.

There is, in addition, one respect in which Grossman's remarks may mislead. The problem of the non-uniqueness of the decomposition does not arise *only* for degenerate density operators *if* we apply the ignorance interpretation to non-orthogonal decompositions as well as to orthogonal ones. So that in one respect Grossman's 'ignorance' defence of the ignorance interpretation may backfire. Park (Park, 1968, pp. 217–9) gives an example of such a density operator which is represented, without cross-terms, in two different non-orthogonal representations. The operator is degenerate in neither representation. In Grossman's account we avoid the non-uniqueness because we do not know that the density operator is degenerate. But then if it is not degenerate, we need to know exactly which pure states represent the mixed subensembles, for otherwise the non-uniqueness may arise again.

This first sort of case is one in which we have a supposed degeneracy leading to the problem of non-uniqueness of decomposition, while we claim to know a privileged representation of the density operator corresponding to the actual composition of the mixture. A rather different case is provided by a mixture prepared by an *incoherent source*, such as the beam of electrons which is boiled off a heated filament. Here again we can represent the mixture by a density operator of the form (#). Grossman's argument covering this case is rather simpler. We can, he says, *choose* to represent the mixture in the (#) for any choice of $\delta$. The choice we *actually* make will generally be determined by the *future experiment* we intend to perform on the mixture. But such choices do not in any way prevent our claiming that each electron in the beam really is in some pure state $|\pm, \delta>$ for some $\delta$.

The argument that the ignorance interpretation leads to inconsistency derives from what is called *the reduction assumption.*[3] The reduction assumption is used in the account of quantum mechanical measurement involving several interacting systems. The conjunction of the reduction assumption and the ignorance interpretation of mixtures leads to paradox in such cases as that discussed by Einstein–Podolsky–Rosen. Proponents of the ignorance interpretation, such as Grossman and Hooker (Hooker, 1972, p. 97ff.) will therefore tend to reject the reduction assumption. In the account offered here, quantum logic enables us to retain both the reduction assumption and (what may be called) a quantum logical ignorance interpretation without encountering paradox.

### *The Reduction Assumption*

The quantum mechanical description of several systems exploits the notion of the tensor product of Hilbert Spaces. Each system has its own Hilbert Space. Omitting niceties, one may say that the elements of the tensor product space consist of the products of the elements of the individual Hilbert Spaces, one element being taken from each. The inner product of two elements of the tensor product space is the product of the inner products from each individual Hilbert Space. And so the tensor product space is a Hilbert Space in its own right. We represent the product operation on the original Hilbert Space elements by $\otimes$.

Suppose we have an ensemble of *pairs of correlated systems* I and II described by the density operator

$$(1) \qquad \hat{\rho}_{\mathrm{I}+\mathrm{II}} = \sum_j \lambda_j \, |j\rangle\langle j|$$

where

$$(2) \qquad |j\rangle = |\alpha_j^{\mathrm{I}}\rangle \otimes |\beta_j^{\mathrm{II}}\rangle .$$

That is, where $\hat{\rho}_{\mathrm{I}+\mathrm{II}}$ describes an ensemble of correlated pairs of systems I and II, pure subensembles of which are described by density operators in the representations $|\alpha_j^{\mathrm{I}}\rangle$ and $|\beta_j^{\mathrm{II}}\rangle$. Note that the appearance of the single variable '$j$' throughout (2) is intended to reflect the *correlation* of I and II.

The problem then is: given a description of correlated pairs in terms of $\hat{\rho}_{\mathrm{I}+\mathrm{II}}$, how can we extract a description of the ensembles of I and II separately?

The answer is given by the *reduction assumption*. According to the reduction assumption, the density operators for ensembles of I and II separately are found by taking the traces of $\hat{\rho}_{\mathrm{I}+\mathrm{II}}$ in any representation of the *other* system in the tensor product space. (The result is, of course, independent of which representation for the other system is chosen.)

For example, (1) is, explicitly

$$\hat{\rho}_{\mathrm{I}+\mathrm{II}} = \sum_j \lambda_j [|\alpha_j^{\mathrm{I}}\rangle \otimes |\beta_j^{\mathrm{II}}\rangle][\langle\beta_j^{\mathrm{II}}| \otimes \langle\alpha_j^{\mathrm{I}}|].$$

So that

$$\hat{\rho}_{\text{I}} = \text{Trace}_{\text{II}}\, \hat{\rho}_{\text{I + II}} .$$

Choosing for convenience the $|\beta_i^{\text{II}}>$ representation

$$\hat{\rho}_{\text{I}} = \sum_i <\beta_i^{\text{II}}|\{\sum_j \lambda_j [|\alpha_j^{\text{I}}> \otimes |\beta_j^{\text{II}}>][<\beta_j^{\text{II}}| \otimes <\alpha_j^{\text{I}}|]\}|\beta_i^{\text{II}}>$$
$$= \sum_j \lambda_j |\alpha_j^{\text{I}}><\alpha_j^{\text{I}}| .$$

Similarly,

$$\hat{\rho}_{\text{II}} = \sum_j \lambda_j |\beta_j^{\text{II}}><\beta_j^{\text{II}}| .$$

A striking feature of this result is that an ensemble of correlated systems may be a *pure case*, while the ensembles for the correlated systems separately may be *mixtures*. This quite non-classical feature of quantum correlated ensembles appears in the Einstein–Podolsky–Rosen paradox in a striking way and plays a part in some of the orthodox resolutions of the paradox, most notably, in that due to Jauch (Jauch, 1968, p. 185ff.). The *quantum logical ensemble interpretation* of quantum mechanics follows Jauch in this. But the importation of quantum logic will allow us to adopt a convention concerning measurement according to which ideal measurements performed on certain mixtures may be said to *reveal* prior membership by an individual system of a pure (sub)ensemble.

## III. QUANTUM ENSEMBLES

The *quantum logical ensemble* interpretation of quantum mechanics embodies the following four claims.

(I) The formalism of quantum mechanics assigns *states* to ensembles and not to individual systems. Quantum mechanics is descriptive of ensembles.

(II) Whenever an individual system $S$ belongs to a mixture, $S$ belongs to some pure (sub)ensemble of the mixture though we are ignorant as to which.

(III) Whenever an individual system $S$ belongs to a pure (sub)ensemble, it has those *properties* corresponding to the subspaces of Hilbert Space containing the statevector describing the pure (sub)ensemble.

(IV) An ideal measurement performed on an individual system belonging to an ensemble $E$ *reveals prior membership* by the individual system of a pure (sub)ensemble $P$ whenever the density operator describing the ensemble $E$ is (diagonally) decomposable into a set of pure (sub)ensembles of which $P$ is a member.

In such a case the measurement is said *not to disturb* the individual system.

Otherwise, the measurement is said *to disturb* the individual system, and does *not* reveal prior membership of a pure subensemble.

(IV) is a *measurement convention.*

We consider two related languages whose underlying logic is quantum logic. One language is descriptive of individual systems, the other of ensembles.

First, the language for individual systems. Let the metalinguistic propositional variables $A$, $B$, $C$ etc range over the elementary statements of the form 'system $s$ has the value of its dynamical variable $D$ contained in some range $\Delta$'. Such statements assign *properties* to the individual system. As an example, consider the elementary statements

$X_{\pm}(s)$: system $s$ has spin-$\pm$ in the $x$-direction;

and similarly

$Z_{\pm}(s)$: system $s$ has spin-$\pm$ in the $z$-direction.

Suppose now that $s$ belongs to an ensemble described by a density operator of the form (#). It follows (II) and (III) that (1), (2) and their conjunction (3) are true

(1) $X_+(s) \vee X_-(s)$

(2) $Z_+(s) \vee Z_-(s)$

(3) $[X_+(s) \vee X_-(s)] \wedge [Z_+(s) \vee Z_-(s)]$ .

However, the distributive expansion of (3) is false, since no system $s$ can, according to the quantum mechanical account of electron spin, belong to pure subensembles of spin-+ in the $x$-direction and spin-+ in the $z$-direction simultaneously, etc. Each of the four statements

(4) $X_{\pm}(s) \wedge Z_{\pm}(s)$

and their disjunction is false.

The distributive expansion, though classically legitimate, is quantum logically invalid. And it is precisely this expansion which leads the ignorance interpretation of mixtures into paradox. We are therefore free to adopt what might be called a *quantum logical ignorance interpretation of mixtures*, according to which:

(QLI²M) Whenever a system $S$ belongs to an ensemble described by a mixed density operator $\hat{\rho} = \sum_j \lambda_j \,|j\rangle\langle j|$ then $S$ belongs to one of the pure (subensembles) described by the density operator $|j\rangle\langle j|$, for some $j$.

The difficulties encountered by the ignorance interpretation of mixtures, and by that interpretation taken in conjunction with the reduction assumption, may all be avoided in the quantum logical description of ensembles, for they all arise, I claim, from illegitimate applications of the distributive expansion.

The language used for the description of ensembles, whose underlying logic is quantum logic, is related to that used for individual systems. Let $E$ be an ensemble. Then corresponding to any elementary statement of the language for individual systems $A(s)$ there is an elementary statement of the language for ensembles $A(E)$ such that

$$A(E) \quad \text{iff} \quad (\forall s)[s \in E \rightarrow A(s)].$$

Thus, for a mixture $E$ described by a density operator of the form (#), the statements (1′)–(3′) corresponding to (1)–(3) are true.

(1′) $X_+(E) \vee X_-(E)$
(2′) $Z_+(E) \vee Z_-(E)$
(3′) $[X_+(E) \vee X_-(E)] \wedge [Z_+(E) \vee Z_-(E)]$

We turn now from the ignorance interpretation of mixtures to what is, on the quantum logical ensemble view, a related question, namely that of the Einstein–Podolsky–Rosen paradox and non-locality.

*EPR, EPR-B and Non-Locality*

In EPR-B we have an ensemble of correlated systems represented by the state-vector (I, II) as above. Its density operator is

$$\hat{\rho}_{\mathrm{I+II}} = |\psi(\mathrm{I, II})\rangle\langle\psi(\mathrm{I, II})|$$

(I, II) is a vector in the tensor product of the Hilbert Spaces for particles I and II. And so the ensemble of correlated systems is a pure case.

But by the reduction assumption, the density operators for the particles I and II represent mixtures:

$$\hat{\rho}_{\mathrm{I}} = \tfrac{1}{2}\,|+,\delta,\mathrm{I}\rangle\langle+,\delta,\mathrm{I}| + \tfrac{1}{2}\,|-,\delta,\mathrm{I}\rangle\langle-,\delta,\mathrm{I}|$$

and similarly

$$\hat{\rho}_{\mathrm{II}} = \tfrac{1}{2}\,|+,\delta,\mathrm{II}\rangle\langle+,\delta,\mathrm{II}| + \tfrac{1}{2}\,|-,\delta,\mathrm{II}\rangle\langle-,\delta,\mathrm{II}|.$$

We have the quite non-classical result that an ensemble of correlated systems may be a pure case, while the subensembles of its component subsystems I and II and both mixtures. This is not peculiar to EPR-B, for EPR exhibits an exactly parallel feature.[4]

$\hat{\rho}_{\mathrm{I}}$ and $\hat{\rho}_{\mathrm{II}}$ are both degenerate density operators, decomposable into pure subensembles of *spin-up* and *spin-down* in any chosen direction $\delta$. We can therefore apply the measurement convention (IV). A spin measurement on a system in ensemble I will yield the result (say) 'up' in a chosen direction $\delta$. The ensemble I + II is a pure ensemble of pairs of correlated systems of oppositely directed spins. The measurement on the system in ensemble I reveals a property of that system, and, indirectly, a property of the correlated system in ensemble II. We can then infer that the correlated system in ensemble II belonged to a pure subensemble of II of oppositely directed spin.

The resolution of the EPR and EPR-B paradoxes becomes, in the quantum logical ensemble interpretation, an application of the ignorance interpretation of mixtures [QLI$^2$M]. No interaction need be assumed to take place between the two correlated systems.

*Non-Locality*

If we imagine, as we may, that the pure ensemble of correlated pairs is already decomposed prior to measurement, what becomes of the non-locality the EPR thought-experiment is usually held to expose?

The answer is that non-locality takes a semantical form: it concerns the truth-values of counterfactual statements we might make about the properties of individual systems.

Imagine a member of ensemble I, and consider the question: "what is the probability that spin measurements on the pure subensemble of I of which the element is a member will yield 'up' in direction $\delta_1$?" Our answer to this question will depend on which direction we actually chose to measure the spin of the particle. Suppose we measure its spin in direction $\delta_2$ and suppose the result is 'up'. Then we say the particle belonged to a pure subensemble of I whose spins were all 'up' in direction $\delta_2$. But suppose we had chosen a different direction again, say $\delta_3$, and suppose the result were again 'up'. Then in both cases we could calculate the probabilities of spin 'up' results for direction $\delta_1$. And we will find these probabilities will in general different. So the truth-values of counter-factual statements about the probabilities of measurement results will depend on the actual measurements we choose to make.

These probabilities are all decidable experimentally. Imagine an experimenter making alternate spin measurements on a sequence of systems taken from I in the directions $\delta_2$ and $\delta_3$, and nothing which give the answer 'up' for each direction. The second experimenter will make spin measurements on the correlated systems in II but this time in the fixed direction $\delta_1$. He too will obtain a sequence of 'up' and 'down', but the first experimenter can tell him which to discard (the ones not correlated to 'up'), and which to associate with each direction $\delta_2$ and $\delta_3$. From this information the second experimenter can calculate the required probabilities. And they will, in general, be unequal.

Finally, a remark about realism. The interpretation of quantum mechanics suggested here is, of course, not realist. The objects which the formalism describes a virtual ensembles, not real physical systems. Dynamical variables do not always take precise values. This might reflect badly on taking the trouble to employ quantum logic. But then the interesting question, to me, is not whether quantum mechanics can be interpreted realistically. The interesting question is rather into what *kind* of anti-realism does quantum mechanics force us?

## NOTES

[1] Van Fraassen (1972), p. 329.

[2] On this see Ballentine (1970), p. 361.

[3] See the discussion in Jauch (1968), pp. 179–185.

[4] Translating EPR into the density operator formalism causes some problems. But, roughly speaking, $\psi(x_I, x_{II})$ is a 'sum' of equally weighted momentum 'eigenstates'. The artificial density operator $|\psi(x_I, x_{II})><\psi(x_I, x_{II})|$ can be thought of as non-denumerably degenerate. The Fourier transform of $\psi(x_I, x_{II})$ allows one to talk in a similar way about position eigenstates, and a corresponding degeneracy in the density operator in the position representation. The artificiality here is due to the appearance of Dirac delta-functions.

## REFERENCES

Ballentine, L. E.: 1970, 'The Statistical Interpretation of Quantum Mechanics', *Rev. Mod. Phys.* **42**, 358.

Bohm, D.: 1951, *Quantum Theory*, Prentice-Hall.

Einstein, A., Podolsky, B., and Rosen N.: 1935, 'Can Quantum Mechanical Description of Reality be considered Complete?', *Phys. Rev.* **47**, 777.

van Fraassen, B.: 1972, 'A Formal Approach to the Philosophy of Science', in *Paradigms and Paradoxes*, R. Colodny (ed.), Pittsburgh, pp. 303–366.

Grossman, N.: 1974, 'The Ignorance Interpretation Defended', *Phil. Sci.* **41**, 333.

Hooker, C.: 1972, 'The Nature of Quantum Mechanical Reality', in *Paradims and Paradoxes, op. cit.* pp. 67–302.

Jauch, J. M.: 1968, *Foundations of Quantum Mechanics*, Addison-Wesley.

Park, J. L.: 1968, 'Nature of Quantum States', *Am. Journ. Phys.* **36**, 211.

Popper, K.: 1971, 'Particle Annihilation and the Argument of Einstein, Podolsky and Rosen', in *Perspectives in Quantum Theory*, Yourgrau W. and van der Merwe A. (eds.), Dover Books, pp. 182–198.

Redhead, M. L. G.: 1981, 'Experimental Tests of the Sum Rule', *Phil. Sci.* **48**, 50.

# NOTES ON CONTRIBUTORS

J. L. MACKIE was a Fellow of University College, Oxford and a Reader in Philosophy at the University of Oxford. Among his books is *The Cement of the Universe*, a study of causation.

JON DORLING is Professor of Philosophy of Science at the University of Amsterdam. Until recently he was Lecturer in the History and Philosophy of Science at Chelsea College, London.

ELIE ZAHAR is Reader in Philosophy at the London School of Economics.

LAWRENCE SKLAR is Professor of Philosophy at the University of Michigan, Ann Arbor. His *Space, Time and Space-Time* was published in 1974.

RICHARD SWINBURNE the editor of this volume, is Professor of Philosophy at the University of Keele. He is the author of several books on the philosophy of science and on the philosophy of religion, including *Space and Time* (second edition, 1981).

RICHARD HEALEY is Assistant Professor of Philosophy at the University of California, Los Angeles. Until recently he was a Research Fellow in the Department of History and Philosophy of Science, University of Cambridge.

W. H. NEWTON-SMITH is a Fellow of Balliol College, Oxford and a Lecturer in Philosophy in the University of Oxford. His *The Structure of Time* was published in 1980.

NANCY CARTWRIGHT is Professor of Philosophy at Stanford University, California.

JEREMY BUTTERFIELD is Assistant Lecturer in Philosophy in The University of Cambridge.

MICHAEL REDHEAD is Lecturer in the History and Philosophy of Science at Chelsea College, London.

PETER GIBBINS is Lecturer in Philosophy in the University of Hull.

# INDEX OF NAMES